THE SCIENCE AND SPIRIT OF SEAWEED

CHRIS ADAIR PHOTO

THE SCIENCE AND SPIRIT OF

Seaweed

Discovering Food, Medicine and Purpose in the Kelp Forests of the Pacific Northwest

AMANDA SWINIMER

HARBOUR PUBLISHING

4 5 6 7 8 — 29 28 27 26 25

HARBOUR PUBLISHING CO. LTD.
P.O. Box 219, Madeira Park, BC, V0N 2H0
www.harbourpublishing.com

FRONT COVER PHOTO by Chris Adair
BACK COVER PHOTO by Jennifer Jellett
BACK COVER ILLUSTRATIONS by Claire Watson
ALL PHOTOGRAPHS by Amanda Swinimer except where otherwise noted.
ALL ILLUSTRATIONS by Claire Watson except where otherwise noted.

EDITED BY Caroline Skelton
INDEXED BY Nicola Goshulak
COVER AND TEXT DESIGN BY Shed Simas / Onça Design
PRINTED AND BOUND IN South Korea
PRINTED on paper certified by the Forest Stewardship Council® with vegetable-based ink

MIX Paper | Supporting responsible forestry FSC® C140526

Harbour Publishing acknowledges the support of the Canada Council for the Arts, the Government of Canada, and the Province of British Columbia through the BC Arts Council.

LIBRARY AND ARCHIVES CANADA CATALOGUING IN PUBLICATION
Title: The science and spirit of seaweed : discovering food, medicine and purpose in the kelp forests of the Pacific Northwest / Amanda Swinimer.
Names: Swinimer, Amanda, author.
Description: Includes bibliographical references and index.
Identifiers: Canadiana (print) 2021023945X | Canadiana (ebook) 20210239603 | ISBN 9781550179613 (softcover) | ISBN 9781550179620 (EPUB)
Subjects: LCSH: Marine algae. | LCSH: Marine algae culture. | LCSH: Marine ecology. | LCSH: Kelp bed ecology. | LCSH: Marine algae as food.
Classification: LCC SH390 .S95 2021 | DDC 338.3/7189—dc23

This book is intended as an educational resource. If you plan to forage for your own seaweed, please get the proper education to ensure it is done ethically, safely and sustainably. Although this book contains general information about the medical benefits of various seaweed species, please consult with a medical professional about any specific treatments.

This book is dedicated with love to my mom and dad, who have always encouraged me to follow my dreams, and to my sister Jen and brother Dan, who have cheered me on every step of the way.

I have felt my life rise
and fall
and rise again to the rhythm of the tide.
I have taken it all out there with me.
The good and the bad,
the pride and the shame,
the fear and the joy,
the sadness and the love,
and I have woven it with the mighty kelps,
the brilliant greens
and the blushing reds.
Together, we have forged a gift of this life I have been given.

TABLE OF CONTENTS

EMMA GEIGER PHOTO

If you don't become the ocean, you'll be seasick every day.

—LEONARD COHEN

CHRIS ADAIR PHOTO

FOREWORD

Seaweeds are probably my biggest life passion. I have eaten, studied and taught about seaweeds for many decades. Seaweeds are my main vegetable, and I am happy to say they have kept me healthy, despite my sometimes not so healthy life choices. As a young student, Amanda was quick to learn what I had to share, and over the years she has become a close friend. She is deeply involved with and in love with the ocean. If you saw her out in our cold North Pacific waters hand harvesting bull kelp, you would know why we call her the Mermaid of the North Pacific. When she first came to one of my seaweed classes, she was thirsty for the knowledge I had to share, and she has continued researching all these years. I am so proud of her work that has now come to fruition in *The Science & Spirit of Seaweed.*

This wonderful book is filled with scientific knowledge, love and spiritual understanding, which is a rare yet refreshing combination. It will guide whoever reads it into the mysteries within the world of seaweeds. Enjoy.

—VERA RØNNINGEN, MSC

JENNIFER JELLETT PHOTO

MAHINA BURLEY PHOTO

INTRODUCTION

"The cure for anything is salt water—sweat, tears, or the sea."

—ISAK DINESEN

Before the First Harvest

I learned about WWOOF, Willing Workers On Organic Farms, in my early twenties, when I was volunteering at a permaculture farm called Finca Ipe on the west coast of Costa Rica. Others who were living and working on the farm, like I was, told me that they had heard about it in the WWOOF catalogue. They told me tales of travelling all over the world by living, eating and volunteering on organic farms.

Back in Canada, I was able to get my hands on a WWOOF catalogue. In the back, on the very last page, there was a listing for one volunteer position with a woman harvesting bull kelp on the south end of Sidney Island. Sidney Island is a small island with a very long, sandy spit at the north end. This part of the island is a park and a favourite place for sailboats to anchor. A seasonal foot ferry to the island runs from the seaside town of Sidney, located on the southeast tip of Vancouver Island. Many people travel to the island for an afternoon picnic or to camp in one of the campsites located near the beach. The rest of the island, which is about seven kilometres long, is grasslands, forest and logging roads. Or at least it was when I was there, more than twenty years ago.

The WWOOF catalogue gave no instructions for contacting this woman, other than a mailing address with a note that mail was only picked up "about once a month." I decided that it would be faster to make a visit in person. In case she accepted me as a volunteer, or in case I got stranded trying to find her, I brought my travel pack, which I spent much of my twenties living out of.

I made it to Sidney and found the ferry terminal on the dock at the end of the main drag. As the ferry pulled away from the dock and civilization on that sunny morning in June, the feeling of pure freedom and undeniable fate swelled in my body and my soul, filling every cell. Whenever I am on the precipice of an unknown journey, I have this feeling: the feeling of infinite possibilities, of leaping off a cliff. The feeling of pure freedom.

I looked over at my dog, Willet, a husky malamute, to share some of the overflow of this amazing feeling. Though the sea breeze was teasing his thick fur, as it was my long hair, his expression was markedly different from mine—it was one of stress, intense focus and stark fear. While the ocean was the truest version of my happy place, it was the ultimate unknown for Willet. He was a landlubber. The two of us would end up living on a sailboat, and while Willet did indeed learn to relax when the boat was docked, the instant we set sail, he would go into survival mode.

After a short but spectacular twenty-five-minute ferry ride, the boat pulled up to a cute little dock and we climbed out. I turned my head around, taking everything in. In the area where we disembarked, there were campsites with tents and families. To my left there was a long spit of sand. It stretched way out into the sea and was dotted with many sailboats. To my right was forest. I started walking south, toward the forest.

Once Willet and I were immersed in the trees, a maze of logging roads appeared as well as evidence of recent logging. We stuck to the road, following the one that seemed the straightest and trying to keep a southerly trajectory. There were so many deer, more than I have ever seen in one area, and Willet quickly gave up trying to chase them.

An hour passed, and then another. It was hot and my pack was heavy, but the excitement of the unknown propelled me onward. Finally, after hours of walking, we saw something. As the aroma of the sea tempted my nose and the sun was beginning its slow descent, we stepped out of the thick forest into an opening that led to the shore and then the ocean. Rows of yellow rope were hung with long, dark green strips of seaweed. A woman walked toward us.

"Hello," I said. "My name is Amanda, and this is my dog, Willet. I have come here to volunteer to harvest seaweed with you."

The young woman replied, "I don't need any help this year because my boyfriend is helping me. But you are welcome to camp on the beach if you like." She gestured to a rocky beach beside the seaweed-drying area and down a slope. "Here is some seaweed for you." She presented me with a small bag of perfectly dried and cut olive-green strips. The label on the bag read "Bull Kelp."

I accepted the gift and Willet and I headed down to the beach she had pointed out. Once I reached the smooth, flat rocks, I dropped my heavy pack with a generous sigh of relief. Although I would not be spending my summer here, I had arrived for the evening after a long and arduous journey.

I was tired and hungry. I loosened the drawstring on my pack and took out a can of beans, my utensils and a grocery bag filled with dog food. Next, I took out a metal bowl and filled it with water, which Willet slurped up gratefully. With my Swiss Army knife, I opened the can of beans and set them down on the rocks, with the fork in the can. I picked up the bag of bull kelp and opened it. As I did this, I looked out at the ocean. About twenty-five metres out, the water was filled with long tubes, big, round bulbs and long, delicate blades of bull kelp. It looked thick enough to walk on.

I took a piece of dried bull kelp out of the bag, the briny smell of the sea entering my nose. I took a taste. An explosion of flavour occurred on the tip of my tongue. Salty and tingling and with a flavour different from anything I had tasted before, a taste so powerful it seemed to crack open an untouched place inside me where it imparted an ancient wisdom. A flavour that demanded a different kind of respect.

I took a piece of dried bull kelp out of the bag, the briny smell of the sea entering my nose. I took a taste. An explosion of flavour occurred on the tip of my tongue. Salty and tingling and with a flavour different from anything I had tasted before, a taste so powerful it seemed to crack open an untouched place inside me where it imparted an ancient wisdom. A flavour that demanded a different kind of respect.

I took a few more bites, savouring the unique taste of the sea, and then I crumbled some up and put it in with my beans. It felt like the most special and sacred gift. I was in awe.

As I ate, a seal appeared in front of me. It looked curiously at Willet and me, then darted down into the water beneath the bull kelp blades, only to pop back up again a moment later. As I sat on the smooth beach rocks, eating my beans and bull kelp and watching this seal play in the bull kelp forest, I felt completely intrigued by the ancient mystery of seaweed. And every cell, every intuition, every breath was gently whispering in my ear to follow my curiosity.

The First Harvest

Several years prior, I had gone on an herb walk through a mountain forest in the Slocan Valley. A young man led a group of us through the forest, cradled on the west side by the Valhalla Range and on the east side by the Selkirks, and talked about the plants we saw—some that you could eat and others that you could make into teas and medicines. I was truly fascinated and felt something awaken in my consciousness.

Growing up in a small suburban city in southern Ontario, I never considered that you could make medicines with the plants growing all around. And apart from visits to a U-pick strawberry farm once a summer, I had never foraged for my own food. That walk transformed me, and I was eager to learn more.

I began eating plants I had learned about and drying leaves and roots to make tea. When I landed on Vancouver Island, I tracked down an herbal class. During one of the sessions, we travelled to Glenora, in the southern interior of Vancouver Island, to visit herbalist Bernice Woolham. On her herb farm, The Hawthorn, she taught us how to make medicines out of herbs that we gathered there. She was a rare teacher, with unique and highly detailed knowledge, and I yearned to learn from her. For over two years, I spent two days per month gathering herbs and making medicines from them and absorbing Bernice's deep knowledge and wisdom. When I met her, I was a young woman. When my apprenticeship ended, I was a mother of one, with one on the way.

Although I was very passionate about wildcrafting and medicine making, the ocean had always been my deepest passion, and I knew that I wanted to build a life and career around it.

One day during my herbal apprenticeship, and about a year after Willet and I had journeyed to Sidney Island, I found three seaweeds listed in a shiny new wildcrafting book I had just bought. With book in hand and Willet bolting out ahead of me, I headed down the trail that led to the beach in front of the small one-bedroom cabin I was renting. I looked at the pictures in the book, then I walked to the low intertidal zone. Lo and behold, I saw one of the kinds of seaweed in the book, and there was lots of it. Like it explained in the book, I cut the seaweed with my Swiss Army knife above the reproductive sporophylls, leaving several inches of the blade above. When I had ten or fifteen long pieces, I headed back up to my cabin. There, I pinned up rows of hemp twine across the length of the living room and hung up my precious seaweed to dry. That day, I began a journey that would weave together my passions, my innate curiosity, my love of place and my values.

I never stopped harvesting seaweed from that day forward. Each day, when the tide was low enough, I would go down to the beach and harvest *Alaria* in front of my cabin. I researched how to start my own business, and I upgraded my drying area from my living room to a converted garage that became my licensed processing facility.

When I began harvesting seaweeds, I knew nothing about their phenomenal medicinal qualities. I just felt a strong sense that I wanted to go deeper into their world, and I felt profoundly fulfilled when I harvested them. I felt empowered, too, because I was able to provide myself and my community with a robust and healthful wild food. But little by little, I discovered that these mysterious

I just felt a strong sense that I wanted to go deeper into their world, and I felt profoundly fulfilled when I harvested them.

"plants of the sea" were among the most nutritional and healthful foods, and that they are essential to the health of the earth.

I kept following this passionate curiosity, and a whole new world opened up. I found myself becoming more connected with my passions, my values and a community that shares them.

Today, my business, Dakini Tidal Wilds, celebrates nearly two decades of providing the highest quality wild seaweeds to people who are passionate about wild, local, healthy and sustainable food. I also teach people about seaweeds, from early elementary grades to college and university students, and I am invited to communities to teach through private venues. I have travelled to international conferences to inspire people about both the global and personal benefits of seaweeds. And best of all, I spend five months of every year wading in the intertidal zone and snorkelling through the kelp forests of the Pacific Northwest, harvesting seaweed.

The Sea

In Japanese, the character for "sea" is a union of "water" and "mother." The sea is the birthplace of life on our planet, and indeed, seawater has a similar composition to amniotic fluid. Mother Sea: a watery womb birthing the rainbow of life.

The ocean's ties to the feminine are not just maternal. Aphrodite, the Greek goddess of love, beauty, passion, sexual pleasure and fertility, was said to have been made from the foam of the sea. The Greek word for sea foam is *aphros*.

The ocean has a mysterious and omnipotent nature. Nautical tales describe tiny dinghies crossing the ocean and mighty ships being swallowed instantly by giant waves, and both are true. Indeed, when you enter the ocean's fluid world, you are at her mercy.

Much of the mystery of the ocean lies in her unknown depths. The deepest place on our planet, the Mariana Trench in the northwest Pacific Basin, lies 11.03 kilometres beneath the surface. The pressure at this depth is more than one thousand times the pressure at sea level. There is no light and no oxygen, the very things that define life on earth, yet bizarre creatures inhabit this most inhospitable of places. According to the National Oceanic and Atmospheric Administration (NOAA), more than 80 per cent of the ocean remains unmapped and unexplored. An estimated 91 per cent of ocean species remain unclassified, yet to be discovered.

Willet and seaweed at low tide in front of the cabin where I harvested my first pieces of *Alaria* in 2002.

The ocean is the lifeblood of our planet. It covers nearly two-thirds of earth's surface and is an average of 3.7 kilometres deep. Our ancient ties to the salty ocean are present in our blood, sweat and tears.

Seaweeds, including kelp, belong to a very special group of organisms known as algae. Algae, specifically cyanobacteria or, more poetically, blue-green algae, are the most ancient living organisms on earth. They remain remarkably similar to their 1.9-billion-year-old ancestors, the first lifeforms on earth, from which it is theorized that all life on our blue planet evolved. These single-celled algae captured starlight from our sun and converted it into energy. Upon creating this energy, they released oxygen. With oxygen, carbon and a usable form of energy, a life-nurturing environment was created on earth. In this way, algae spun the web of life we see today.

An astounding amount of global biodiversity and biomass exists in the oceans. Algae, in addition to producing an estimated 50 to 80 per cent of the world's oxygen, are the primary producers—that is, the *foundational nourishment*—for this incredible array of life.

Over ninety nutritional elements have been found in algae, making them the world's most nutrient-dense food source. They contain concentrated vitamins, minerals, essential fatty acids, protein, fibre and biologically active compounds not found anywhere else in nature. Peoples whose traditional diet includes seaweeds boast incredible longevity, resistance to chronic and degenerative diseases and a vital constitution.

The relationship between people and seaweed dates back to at least 14,500 years ago, according to archaeological finds in Monte Verde, Chile, which verified the use of seaweeds as food. Out of nine species uncovered, one species, a kelp, is still sold today in Chilean markets.

In Japan, archaeological finds show evidence of the harvesting and eating of seaweeds from as early as 6000 BCE. In Korea, two species of brown seaweeds in the genera *Hijiki* and *Undaria* have been discovered in fossilized meals from 10,000 years ago.

There is not only a long tradition of people eating seaweeds, but an almost equally long tradition of using them medicinally. This is evidenced by their presence in the Ebers papyrus, the oldest preserved medical text in existence. This ancient Egyptian text, dated 1536 BCE, recommends a seaweed believed to be in the *Sargassum* genus of brown seaweeds for the treatment of breast cancer.

I live in the Pacific Northwest. Here we have the greatest cold-water seaweed diversity in the world, with an estimated 650 species. The Pacific Northwest, where the coastal mountains and temperate rainforest converge with the underwater kelp forests of the nutrient-rich emerald sea, nurtures an abundance of life and is grand in its beauty. Thirty species of kelp thrive here in the cold waters, more species than anywhere else on earth. For one who loves seaweed, especially one whose passion within the seaweed world is kelp, it is a real-life paradise.

I have spent most of my adult life among seaweeds. Following my deep childhood passion for the ocean and a healthy dose of curiosity led me to this life, and the ancient beauty of seaweeds has kept me here. Rare and treasured experiences I have had while harvesting seaweed combined with nearly two decades of teaching inspired me to write this book. Seaweeds are my muse.

Seaweeds are shrouded in wonderful mystery. Part plant, part animal and at times fungi-like, they have been revered by coastal cultures the world over as food, medicine, ceremonial props and the subject of myths. Currently, seaweeds are touted as part of the solution to the problems of climate change, ocean acidification and pollution. When I unknowingly set upon my journey with seaweeds on that spring day in 2002, harvesting *Alaria* on the beach in front of my one-bedroom cabin, I knew of none of these grand things. I just felt an intuitive

EMMA GEIGER PHOTO

calling, one that has run deeper with wonder than I ever could have imagined.

This book is my love song to seaweed. In it, I share what I have learned so far, having been immersed in their world for the past two decades. I offer information about individual species, summarize the wealth of healing benefits from both traditional knowledge and current research and share delicious recipes from chefs who make the seaweed I harvest into stunning gastronomic art. Last but not least, I share real experiences from my journey and life among the mysterious and beautiful seaweeds.

I invite you, reader, on an underwater journey into their world, and I hope that this book will impart knowledge and awareness and inspire a celebration of seaweed.

A Note about Sources

I have referred to many sources over the years to learn about the tremendous nutrient content of edible seaweeds, some of which are listed in the Recommended Reading section (page 223). *Pacific Seaweeds* and *Irish Seaweed Kitchen*, both included in the Recommended Reading section, have tables that show specific amounts of nutrients in selected edible seaweeds.

Day 1, Harvest Season #18

I am so fortunate to do what I do. I am so grateful to do what I do.

Being out in the ocean harvesting seaweed is profoundly healing for me on a physical, emotional and spiritual level.

Being able to work at a job that has minimal impact on the environment is a necessary blessing, both for my health and the health of Mother Earth.

Providing people with what I deeply believe is one of the most health-supporting living organisms on the planet gives me a sense of purpose that I need in order to feel whole.

I am so happy and grateful that seaweed and I found each other, and that I had the faith and courage to follow my instinct, my heart and my soul.

That's all I have time to write for now. I am off to harvest.

JENNIFER JELLETT PHOTO

ALGAE AND SEAWEED: A DEFINITION

Algae is a comprehensive term that includes both microalgae and macroalgae—an incredibly diverse group of organisms. The word "seaweed" is a colloquial term for marine macroalgae only.

Microalgae are single-celled microscopic organisms that photosynthesize. They can have either eukaryotic cells, which have membrane-bound organelles such as mitochondria and nuclei, or prokaryotic cells, which lack these membrane-bound organelles.

Those with prokaryotic cells are essentially photosynthesizing bacteria. Bacteria, microalgae and a group of organisms known as Archaea are the only groups of organisms that have prokaryotic cells. Every other living thing on earth—from sponges to trees to spiders, fungi, mosses, humans and macroalgae—have eukaryotic cells.

Macroalgae are mostly multicellular organisms that have eukaryotic cells and photosynthesize. They can be further divided across two different kingdoms into three main groups: the brown seaweeds (class Phaeophyceae), the red seaweeds (phylum Rhodophyta) and the green seaweeds (phylum Chlorophyta). They resemble plants but are more ancient and hence primitive.

The plant-like body of a seaweed is called the thallus. It usually consists of a leaf-like part called the blade or frond, a stem-like part called a stipe and a root-like part called a holdfast. Unlike the roots of a plant, the sole purpose of the holdfast is to anchor the seaweed, most often to a rock. Some holdfasts are made of long narrow projections called haptera, which look very much like the roots of a plant. Some seaweeds have a discoid holdfast that looks like a suction cup.

Seaweeds do not have flowers, cones or complex transport systems for nutrients, though some large kelps have primitive ones. Seaweeds absorb their required nutrients directly from the water that surrounds them. Some seaweeds reproduce asexually, through fragmentation or the production of spores; some reproduce sexually, through the production of eggs and sperm; and some alternate between asexual and sexual reproduction. Seaweeds have an amazing diversity of forms and structures, or morphologies, from the giant kelp to the free-floating *Sargassum* to the intricate coralline algae. Some species, such as *Mastocarpus* spp., have two completely different morphologies at different life stages. In some species, such as dulse, the females are much smaller than the males, nearly microscopic.

CHLOROPHYTA: THE GREEN SEAWEEDS

Chloro comes from the Greek word for "green." The green seaweeds, commonly referred to as "the chlorophytes," belong to kingdom Plantae and contain approximately six thousand species globally. Most green macroalgae live in fresh water, but they can live in almost any habitat on earth, including such inhospitable places as deserts, alpine snow, the Arctic and hydrothermal vents. Like land plants, their dominant photosynthetic pigment is chlorophyll, which gives them their green colour.

RHODOPHYTA: THE RED SEAWEEDS

Rhodo comes from the Greek word for "rose." The red seaweeds are commonly referred to as "the rhodophytes" and belong to kingdom Plantae. The red seaweeds contain the highest number of species, with an estimated seven thousand globally. In the book *Pacific Seaweeds: A Guide to Common Seaweeds of the West Coast*, BC authors Louis D. Druehl and Bridgette E. Clarkston estimate that approximately four hundred of our estimated six hundred fifty local species are red seaweeds. The dominant photosynthetic pigment in red seaweeds is phycoerythrin, and their colour can range from deep wine red to rusty red, and from bright purplish pink to the pastel pinks of the coralline algae. Carrageenan and agar, gelling agents found in some species of red seaweeds, are used commercially as thickeners and stabilizers.

PHAEOPHYCEAE: THE BROWN SEAWEEDS

Phaeo comes from the Greek word for "dusky" or "dark." Commonly referred to as "the phaeophytes," brown seaweeds belong to kingdom Chromista. Unlike the rhodophytes and chlorophytes, the phaeophytes are the only group of macroalgae that share their kingdom with microalgae, specifically unicellular diatoms. There are an estimated two thousand species of brown seaweeds globally. Fucoxanthin is the dominant photosynthetic pigment, giving many brown seaweeds an almost golden hue. Brown seaweeds contain alginate, a natural gum, and fucoidan, a sulphated polysaccharide used as a health-supporting supplement. Nearly all brown seaweeds are marine, and most live in colder waters.

EMMA GEIGER PHOTO

Kelp are brown seaweeds belonging to the order Laminariales. They are large marine seaweeds that grow in the low intertidal and subtidal zones. The majority of scientific studies on seaweeds and human health have focused on kelp and brown seaweeds in the order Fucales. The Pacific Northwest has the richest diversity of kelp on the planet, with thirty described species in eighteen genera, including the largest seaweeds on earth.

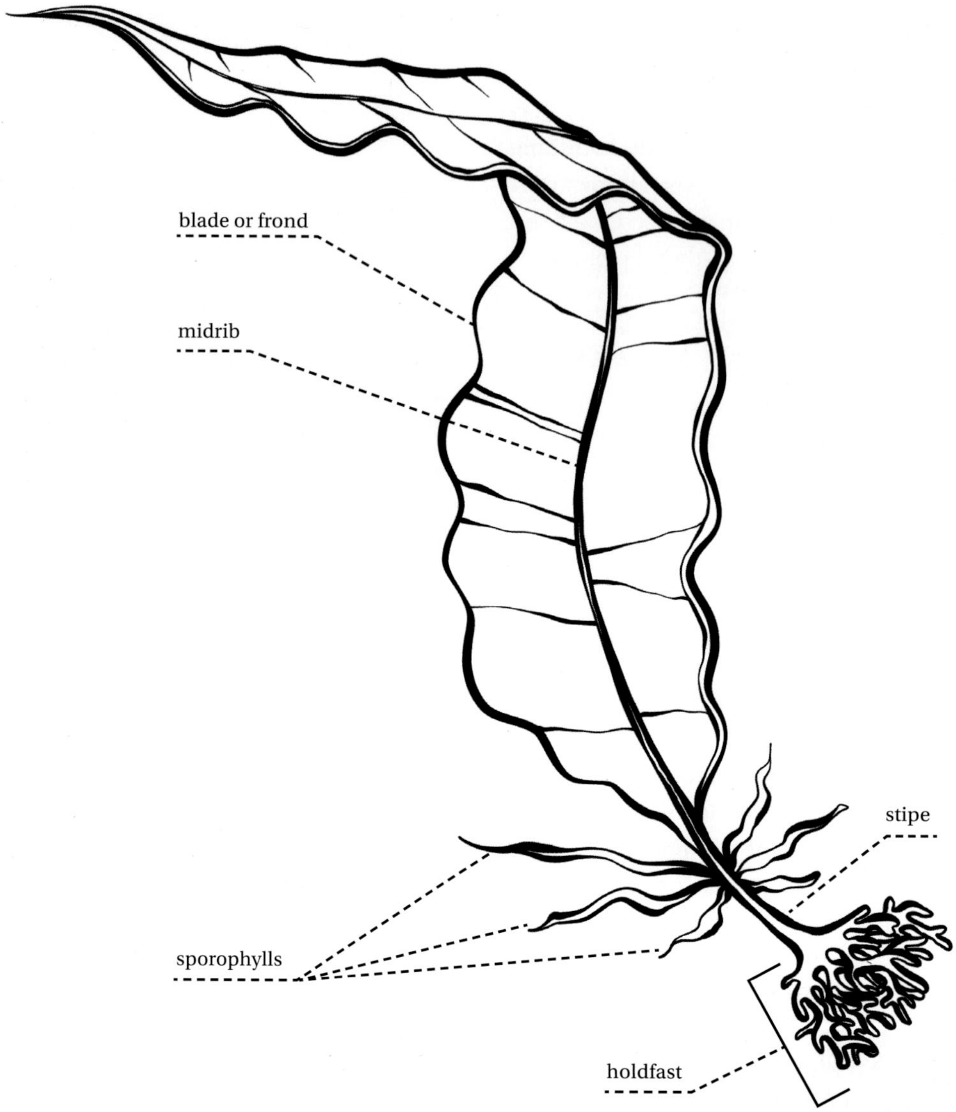

A Quick and Simple Definition of Seaweeds

- "Macroalgae" is the formal name for seaweeds
- Seaweeds resemble plants but are more ancient and primitive
- Seaweeds photosynthesize (although a handful of parasitic reds do not)
- There are three main groups of seaweeds: the browns, the reds and the greens
- Kelp are large brown seaweeds

HARVESTING SEAWEED SUSTAINABLY

Seaweeds are among the fastest-growing organisms on the planet and are very regenerative *if* they are harvested properly. The number one rule for harvesting sustainably is: do not rip the seaweed from the rock. Always leave some of the seaweed attached and harvest by pruning. Here are some key considerations to ensure a sustainable harvest.

Education

For seaweeds to regrow, they not only need to be harvested by leaving some of the seaweed attached to the rock, but they also need to be cut in the right place. The best place to cut the seaweed will depend on the species. Most seaweeds will not regenerate if their stipe (stem-like structure) is cut. In many seaweeds, the new growth happens in the part of the blade (leaf-like structure) closest to the stipe. Leaving this part intact and leaving several inches (or more for large kelp) of the blade is important for the seaweed to be able to regenerate. Furthermore, many seaweeds have separate reproductive structures that need to be left intact to ensure they can reproduce. Still other seaweeds are long-lived perennials and require extra consideration. It is also important to know the ecological role of seaweeds in areas where you are planning to harvest. Take a class and learn about the specific seaweeds you want to wild harvest.

In addition, depending on where you are harvesting, a licence may be required. If you are harvesting in the traditional territory of a First Nation, you need to obtain permission from the nation to harvest there.

Abundance

Only harvest a particular kind of seaweed if there is an abundance of it growing in the area you are planning to harvest from.

Only Take What You Need and Can Process

Seaweed takes an *enormous* amount of space to dry. Ensure that you only harvest what you *need* and what you will be able to dry or process in your space.

Harvesting by Salvaging

Harvesting by salvaging is a great way to be sustainable. Where I live, we routinely get big winds and ocean swell, which rips up healthy seaweeds and deposits them on the beach like treasures for me to find. To determine if the seaweed is still fresh and of adequate quality to be consumed or used in the bath, use your senses. It should smell briny and fresh, it should not be slimy to the touch but should feel lubricated and gelatinous, and the colour should look uniform and vivid. For elusive species, or those found only in deep water, salvaging is often the only way to harvest them.

Reciprocity

Practising reciprocity is essential to sustainability. In healthy ecosystems, there is a balance of give and take among all of the living organisms. When you harvest seaweeds, you become a member of their ecosystem. The organisms that make up an

Listening to the Ecosystem

As I was preparing to harvest bull kelp one day, many years ago, I had a very strong sense that it was not an appropriate day to be "taking." I second-guessed myself and headed out to harvest anyway. As I was walking out of the ocean with a heavy bag full of bull kelp, a wave picked me up and smashed my masked face into the sand. I picked myself up slowly, feeling shocked and disoriented, and began looking around for my bag of bull kelp. Although I searched the area for some time, I never found it. Since then, I take a moment before harvesting to listen to the ecosystem. I ask permission. When I am finished harvesting, I look out at the kelp forest or the intertidal seaweed garden and I say, "Thank you."

Students learn to identify local seaweeds, the health benefits of seaweeds, how to cook with seaweeds and use them in the bath, the biology and ecology of seaweeds and sustainable harvesting techniques at a seaweed workshop.

CHRIS ADAIR PHOTO

ecosystem function together as a whole, and thus it is essential to listen to the ecosystem and be sensitive to its needs. What can you give back to an ecosystem after taking something, in order to restore balance? Giving back is personal and can be done in many ways. Taking action on local threats to the ecosystem where I harvest and global threats to the larger ocean ecosystem are ways in which I choose to give back. Sharing the love and wonder you have for an ecosystem with your community is a great way to engage people and grow support to protect the health and integrity of local ecosystems.

Commercial Harvest

In the province of British Columbia where I work and live, a commercial seaweed harvester is legally obligated to harvest by pruning and is only allowed to harvest by hand, cutting the seaweed a minimum distance from the holdfast or stipe, which varies depending on the species. As a commercial harvester, I feel very fortunate that our local regulations help to ensure that harvesting is done sustainably.

An abundance of colourful, healthy seaweeds growing in the intertidal zone.

Where to Harvest

Seaweeds that are being harvested for food should come from clean water. Urban centres and areas near sewage outfalls are not safe places to harvest. Since seaweeds are masters at concentrating the minerals in the water in which they grow, they should be harvested far away from industrial areas such as pulp and paper mills, mines, shipyards and other industrial sites that discharge chemicals or heavy metals into the ocean or upland waterways. Areas with strong tidal currents are better than areas with more stagnant water, as the currents regularly bring water in from the open ocean and flush the old water out.

CHLOROPHYTA

THE GREEN SEAWEEDS

"A lot of people attack the sea, I make love to it."

—JACQUES-YVES COUSTEAU

Sea Lettuce and Dark Sea Lettuce

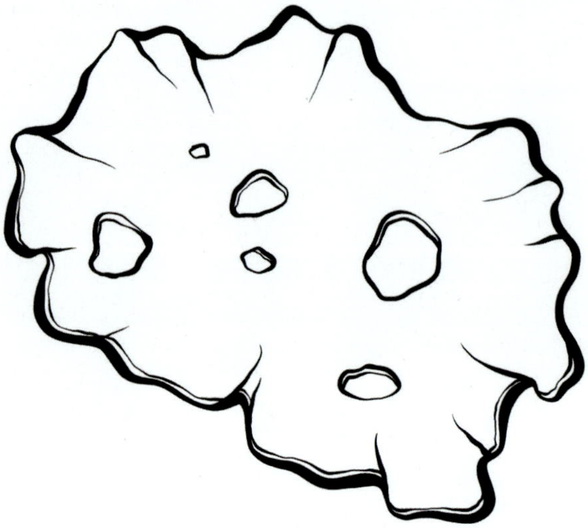

Ulva spp. and _Ulvaria_ spp.

Ulva is Latin for "grass-like aquatic plant."

Sea lettuce exposed at low tide with a blood star.

KINGDOM: Plantae
PHYLUM: Chlorophyta
CLASS: Ulvophyceae
ORDER: Ulvales
FAMILY: Ulvaceae

DESCRIPTION

There are an estimated 128 species of *Ulva* worldwide and about three species of *Ulvaria*, known by the common names "sea lettuce" and "dark sea lettuce" respectively. In *Pacific Seaweeds*, Druehl and Clarkston estimate there are fifteen West Coast species of *Ulva*—eight bladed forms and seven cylindrical forms—and two local species of *Ulvaria*. Distinguishing between fresh *Ulva* species and *Ulvaria* species without a microscope is difficult if not impossible. *Ulva* species are two cells thick while *Ulvaria* species are one cell thick. Both are edible but there is one significant difference: *Ulvaria* contains significant amounts of dopamine (up to 5 per cent dry weight). Bladed species of sea lettuce and dark sea lettuce typically grow up to thirty centimetres in length but some species of *Ulva* can grow up to one metre in sheltered waters. The blades are flat and are a brilliant emerald green that often fades to white at the edges, especially in older specimens. Some morphologies have holes in them reminiscent of Swiss cheese, and these used to be identified as *Ulva fenestra* (*fenestra* means "window" in Latin). However, it is now known that *Ulva* can have a stunning diversity of morphologies within a single species and this species name has since been dropped. The cylindrical forms, every bit as tasty as the bladed forms but much more laborious to process, are usually twenty centimetres or shorter but can grow up to fifty centimetres in length.

RANGE

Local species of *Ulvaria* grow in the intertidal and subtidal zones from Alaska to Mexico.

Local species of *Ulva* grow in the intertidal and subtidal zones from Alaska to Mexico. This genus is extremely prolific and can be found on most of the world's coastlines.

Sea lettuce with giant green anemones. *Anthopleura xanthogrammica*, the giant green anemone, lives in a symbiotic relationship with microalgae, which is the source of its beautiful blue-green colour.

NUTRITIONAL PROPERTIES

Sea lettuce is an excellent source of vitamins A, B, C and E and minerals including iron, calcium, manganese, potassium and silica and is one of the richest known sources of iron. It is an excellent source of fibre, protein and essential fatty acids including omega-3s. *Ulvaria* species contain dopamine, a hormone and neurotransmitter in the brain associated with reward-motivated behaviour and the sensation of pleasure.

AS A HEALING AID

Sea lettuce (including dark sea lettuce) is a traditional remedy for various health conditions: it is consumed to prevent colds and flus, it is applied topically for burns and inflammation and it is made into a hot tea to treat intestinal worms.

EMMA GEIGER PHOTO

Ground sea lettuce and bull kelp: my favourite seasoning for popcorn.

HARVESTING

I harvest sea lettuce at low tide. I try to find seaweeds still covered in water to avoid meticulous rinsing after harvesting. Sea lettuce (including dark sea lettuce) is prone to collecting univalves, isopods, sand and even little shore crabs so it needs to be well rinsed. I leave a few inches attached to the rock and cut.

It is important to know that sea lettuce is a very opportunistic species, meaning that it will outcompete other seaweeds in areas where there is an abundance of nutrients. For example, on rocky beaches near sewage outfalls and in areas of agricultural runoff, you will often see plenty of sea lettuce growing with few, if any, other species. This is a good indicator that you should not harvest in that area. In extreme cases, sea lettuce will grow so abundantly that it will use up most of the available nutrients and shade out other seaweeds and sea grasses. As these other species and then the sea lettuce itself begins to die, bacteria that decompose the dead seaweed use up an enormous amount of oxygen. This creates anaerobic conditions, which drives off and can kill other organisms, including invertebrates and fish. Rotting sea lettuce produces noxious hydrogen sulphide gas, which has a stench that has become infamous in places such as Venice, Italy. There are unconfirmed accounts, both historical and recent, that cows, pigs, dogs and even humans have died due to exposure to hydrogen sulphide from decaying sea lettuce.

CULINARY

Sea lettuce has a fresh "green" taste that is distinct from other seaweeds. Its delicate texture makes it a lovely addition to salads. When dried it has a strong, pungent odour and flavour. Dried sea lettuce is a great snack: right out of the jar, mixed with nuts, or sprinkled on popcorn. Chefs love the flash of electric green it adds to a dish, and my friend and colleague Christine Hopkins makes a beautiful pesto with it.

SPA

Like all edible seaweeds, sea lettuce can impart its oceanic healing benefits through skin absorption. I personally reserve my supply for kitchen use, as sea lettuce is quite small and very tasty.

APRIL 2019

A Seaweed by Any Other Name

To follow your soul's desire and take that leap of faith over and over again, without seeing where it will take you or where you will land, is trusting the wisdom of the wild, feminine force within.

I originally saw the word "dakini" in the book *Women Who Run With the Wolves*, by Dr. Clarissa Pinkola Estés. I read this book while travelling through Central America and it changed me. It helped me to understand, on a deep level, things I intuitively felt.

The word appears near the beginning of the book, where Dr. Estés explains that many cultures have their own word for a wise, wild and dancing spirit unique to women. "Dakini" is the Tibetan word for such spirit.

Somehow, when I was coming up with all the potential names for my new business back in 2003, this word made the cut. And now, eighteen years later, I realize I couldn't have chosen a more perfect name.

Because to do something for eighteen years and still love it as deeply—more deeply—than when you started, involves a beautiful and free-spirited kind of dance. To follow your soul's desire and take that leap of faith over and over again, without seeing where it will take you or where you will land, is trusting the wisdom of the wild, feminine force within.

Other Common Green Seaweeds of the Pacific Northwest

ACROSIPHONIA COALITA >

Acrosiphonia species look like stringy green hair.

< CODIUM FRAGILE

Codium species are eaten in some regions and have traditionally been made into a tea to treat intestinal worms. They have a spongy texture.

PRASIOLA SPP. >

Growing above the high tide line, *Prasiola* species look like bright green moss.

CHRIS ADAIR PHOTO

RHODOPHYTA

THE RED SEAWEEDS

"The voice of the sea speaks to the soul. The touch of the sea is sensuous, enfolding the body in its soft, close embrace."

—KATE CHOPIN, *THE AWAKENING*

Nori

Black nori.

Pyropia abbottiae* and *Pyropia nereocystis

Pyropia abbottiae was named for the Hawaiian phycologist Isabella Aiona Abbott; *Pyropia nereocystis* was named for its association with growing on the stipes of bull kelp, *Nereocystis luetkeana*.

KINGDOM: Plantae
PHYLUM: Rhodophyta
CLASS: Bangiophyceae
ORDER: Bangiales
FAMILY: Bangiaceae

DESCRIPTION

Nori is one of the most popular and widely cultivated edible seaweeds in the world, and it has become popularized in the West by its use in sushi. Most nori used in sushi restaurants and sold in grocery stores as "seaweed snacks" are one of two species of *Pyropia*, cultivated extensively in Japan, China and Korea. Known as "laver" in Scotland and England and as "sleabhac" in Ireland, *Pyropia* has been eaten there for centuries. In the Pacific Northwest, two species of nori, *Pyropia abbottiae* and *Pyropia nereocystis*, are excellent edible seaweeds. *Pyropia abbottiae* is still harvested by several coastal First Nations as a traditional food source. Each species of *Pyropia* has a slightly different flavour, but they all share the same unique texture, being just one cell thick.

Pyropia abbottiae, known locally as "black nori" or "black weed," is a small (up to approximately fifty centimetres long), translucent, dark green seaweed that appears black when dried out on the rocks at low tide. It is attached by an almost indiscernible discoid holdfast. It grows in the mid to high intertidal zone on exposed and semi-exposed rocky shores. Its edges are frilled.

Pyropia nereocystis, known locally as "red nori," is a subtidal species and often grows on the stipes of *Nereocystis luetkeana* (bull kelp). It is quite large for the genus and can grow to a metre long. It is composed of a single flat blade, lacking the frills of *Pyropia abbottiae*, and is a pale shade of burgundy. Like *Pyropia abbottiae*, it is only one cell thick, giving it a translucent appearance, and it too has a tiny discoid holdfast.

RANGE

Both species are found on the west coast of North America from Alaska to Northern California.

NUTRITIONAL PROPERTIES

Pyropia species are the highest in protein of all edible seaweeds. Depending on the species,

Red nori. EMMA GEIGER PHOTO

location and season, they contain approximately 30 to 50 per cent protein, most of which (an estimated 75 per cent) is digestible. They contain approximately 3 per cent lipids by dry weight, of which 50 per cent are essential fatty acids, including omega-3s. *Pyropia* species are rich in vitamins A, B-complex (including B12), C, D, E and K, and they are an abundant source of essential minerals including calcium, iron, iodine, phosphorus, potassium, manganese, copper, selenium and zinc.

AS A HEALING AID

Pyropia species contain omega-3 fatty acids, including eicosapentaenoic acid, which studies indicate decreases the risk of heart disease and cancer, reduces inflammation, helps treat depression and symptoms of menopause and is important for healthy brain function. Algae (including seaweeds) and oily fish are the only known sources of eicosapentaenoic acid.

Pyropia species are effective at lowering blood sugar and can be helpful for people with type 2 diabetes and people at risk for type 2 diabetes. Research suggests that regular consumption can provide protection from UV light and support eye health.

An estimated 30 per cent by dry weight of *Pyropia* species is a sulphated polysaccharide and soluble fibre called "porphyran." Scientific studies report anti-tumour properties of porphyran, including selective cell apoptosis on isolated human cancer cells in vitro (see Seaweed and Cancer, page 134). Research also indicates that porphyran has antiallergic properties.

Nori has a long history of traditional use as a healing aid. This includes its use to cleanse and tone the cardiovascular system (specifically by aiding circulation and decreasing cholesterol and blood pressure); to protect the liver; to lower blood sugar and protect against diabetes; to encourage the dissolution of cysts; to prevent and treat goitre; and to prevent and treat beriberi, a disorder caused by thiamine (vitamin B1) deficiency. It has also been used as an antiviral and antibacterial (specifically to treat urinary tract infections), an anti-inflammatory and a tumour preventative.

SPA

Nori is very nourishing to the skin, helping to improve tone and impart moisture.

HARVESTING

Pyropia is by far the most labour-intensive seaweed I harvest. Because it is harvested on dry land, it needs to be meticulously rinsed in clean ocean water. If you rinse in fresh water, the cells' membranes burst due to the difference in osmotic pressure between fresh water and salt water, and the seaweed will begin the process of deterioration. Even though this seaweed will not rot if properly dried, because the salts are a great preservative, the seaweed will not be as vital. If you are planning to consume it right away, then washing in fresh water is fine. Nori also takes an enormous amount of space to dry and hours of time to spread. It is plenty worth it though: nori is a delicious seaweed.

Pyropia abbottiae

Pyropia abbottiae, or black nori, is the only seaweed I harvest with scisors as opposed to a knife. Its rubbery texture when fresh is hard to cut through with a knife and much easier with scissors.

The holdfast is tiny and does not grip very strongly, like a kelp would, and therefore harvesting must be done gently and carefully to ensure that the holdfast remains attached. I leave a good three or four inches (seven to ten centimetres) attached to the rock and, carefully holding the seaweed in one hand, I cut with the other. I always leave some seaweeds on each rock untouched. After I have approximately four litres of seaweed, I head from the high intertidal zone to the water's edge and begin the process of meticulous rinsing. Because black nori is harvested on dry land, it is quite sandy. Also, creatures such as isopods and univalves love to hide in the seaweed's frilly folds. This relatively small amount of cut nori will fill all four of the big screens in my shop when spread to dry.

The spreading can take hours and defines the term "labour of love." The texture of nori is sticky in the way that plastic cling wrap is, so once you get one corner of it down, the other corner can roll up. You can just hang it to dry, and it will dry all rolled up, but since my daughters and I love to toast it in the oven, I take the time to spread it. Because it is only one cell thick and weighs so little, you will be left with very little final product. But that first bite of savoury, salty deliciousness instantly makes it all worthwhile.

Pyropia abbottiae, black nori, dries into a multitude of interesting shapes on the screen.

Pyropia nereocystis

I harvest all of my *Pyropia nereocystis*, or red nori, by a technique I call salvaging. In this method of harvesting, I look for seaweed that has washed up on the beach, or is floating near the shore, that has been detached but is still very fresh and vital. This is a highly sustainable way to harvest, and for some deeper water and elusive seaweeds, it is the only way I harvest them. I have a secret beach where I find lots of red nori washed up in the summer. When I find pieces lying in the sand, I take them back to the water to rinse them. When they are floating, I rinse them as I go. I dry them the same way as black nori, but because they are so much larger and lack the frilly edges, the spreading takes less time.

Toasted black nori.

Mother of the Sea

Nori used to be called "gambler's grass" by Japanese nori farmers because some years they would have an abundant harvest and other years almost nothing. Farmers knew that if you cut the seaweed above the holdfast you could get another crop, but the harvest varied dramatically and no one could figure out why. Kathleen Drew-Baker was a British phycologist who researched the life cycle of *Pyropia* (then called *Porphyra*). She was forced to leave her job as a researcher and professor at the University of California in 1928 after she married, due to the university's policy of not employing married women. She continued her work privately and discovered that the life cycle of nori includes a microscopic diploid phase (in which it possesses two sets of chromosomes). During that phase, it requires shells to grow on. The seaweed we recognize as nori is the macroscopic haploid phase (in which it possesses a single set of chromosomes) of the species. Japanese nori farmers learned of her research, added shells to their farms and credit her with saving the industry. In 1963, a statue was erected in her honour in Uto, Japan, where she is known as Mother of the Sea.

CULINARY

Because it is only one cell thick, nori has a unique texture. It is most often eaten toasted, such as in sushi and seaweed snacks. When it is fresh, it has a rubbery texture that many find unpalatable, but light toasting breaks down the cell walls and then the seaweed melts in the mouth. The nori sheets used to make sushi are made in a similar way to paper: the seaweed is shredded and then pressed into sheets and toasted. My favourite way to eat nori is tossed in either sesame oil or olive oil and lightly toasted. Laverbread, a traditional English dish, is made by boiling the nori for several hours and making it into a paste.

It is fun to taste red and black nori side by side. Though their texture is the same, they differ in flavour. The red nori has a stronger, brinier flavour, in my opinion, than the milder black. Friends and family who have tried them like this all have a personal favourite, but overall, the black is more popular.

Pyropia nereocystis, red nori, drying on a screen.

JUNE 2018

My Birthday Bear

It is my birthday. The low tide is in the late morning, allowing me to have a pleasant and relaxing start to my day. My daughters wish me a happy birthday upon waking, and the sun is shining bright. It's true that I have to work on my birthday, but when you love what you do, that can be a good thing.

I feel anticipation at the thought of being in my seaweed garden and wonder what kind of gift it might offer today.

Kelly, my apprentice this *Alaria* season, wanders down from the camper on the north end of my property, and we slowly begin suiting up. I feel anticipation at the thought of being in my seaweed garden and wonder what kind of gift it might offer today.

We drive the seven kilometres to our harvest spot, then begin the walk with our wheelbarrow down the 150-metre trail that runs parallel to the creek. By the time my birthday rolls around each year, the end of the trail has become a tunnel of salmonberry bushes. The moment you step out from it onto the washed-up driftwood and smooth rocks of the high tide line, the vista is stunning.

We pop out of the bushes and onto the beach. After the hot, stagnant inland hike, a gentle ocean breeze laps at our foreheads.

To our right the mouth of the creek joins the Pacific Ocean. At this juncture, there are often small, fast-breaking whitecaps or perfectly formed swells—miniature versions of the rideable swells of winter.

Farther west, there is a peninsula made of large boulders, completely dry at low tide, and the extensive seaweed gardens where we will harvest. *Alaria*, *Egregia*, bull kelp, *Costaria*, *Cymathaere*, *Desmarestia*, sea cabbage, rainbow, *Prionitis*, *Fucus*, all the powder pinks of the corallines and large swaths of nori, dulse, sea lettuce and so many more await us. In the distance, as far west as one can see on land, the Sheringham Point Lighthouse stands guard, protecting sailors as it has since 1912.

I feel the same as I have for the past sixteen years, arriving at this spot: as if I have entered a different world. I take in a generous breath of the salty air.

Kelly and I make our way over the uneven rocks to the peninsula of boulders and lay our bags down on a rock—the same rock I have been laying my harvest bags down on for years. As I begin to pull on my glove, I spot something down the beach. It isn't moving, but the black is too rich to be a log, or really anything other than a . . . bear.

And then I see a tiny movement. A moment later, two small bear cubs pop out of the woods close to the beach where their big mama is standing. She appears preoccupied with something in the beach rocks, digging and sniffing. The cubs check in with her and head back into the woods, just a few feet away.

"Look, Kelly," I say, pointing toward the bear. "A bear. On my birthday. Awesome!"

We watch as the mama bear continues to be completely preoccupied with whatever is in the rocks. Every so often, her two cubs pop out, check in with Mom and then disappear back into the woods. The mama bear barely looks up from whatever has her attention.

The bear is far down the beach and not showing any interest in us, so we begin our harvest.

Wading out carefully over the slippery, seaweed-covered rocks of low tide, we each find a nice patch of healthy *Alaria* in waist-deep water.

As I lift the first *Alaria* out of the water with my gloved hand, I take my bare hand to the frond in front and gently follow my way toward the seaweed's holdfast. When I feel an ever-so-gentle resistance, I lean down and cut, leaving several feet of the seaweed to regenerate. By this time of year, it is not uncommon to harvest from a seaweed we have already harvested from earlier in the season. Seaweeds are amazingly regenerative and among the fastest-growing organisms on the planet.

With my bare hand, I give the seaweed a scrub in the ocean water along its midrib, cutting at the top of the seaweed where it starts to look ratty, and where the cells naturally begin to die off. I then tuck the seaweed into my bag and move on to find another beautiful one.

Kelly and I find ourselves distracted, looking up often to see if the mama bear is still there with her two cubs.

We each fill three bags per harvest session. When one bag is full, we carefully "skate" over the slippery, seaweed-covered rocks with our thick wetsuit booties, heavy bags of seaweed humped over our shoulders.

The bear sticks around for our first two bags before disappearing with her cubs into the rainforest lining the beach.

I feel awash with gratitude. Harvesting *Alaria* alongside a mama bear and her two cubs: what a birthday gift. I feel privileged to be considered a natural part of the ecosystem, someone around whom local creatures feel they can go about their usual business.

It is yet another gift from spending time with the seaweed. As I push the wheelbarrow toward the salmonberry tunnel and the path home, I give one last glance over my shoulder and think to myself, "I have the most amazing office in the world."

Pacific Dulse

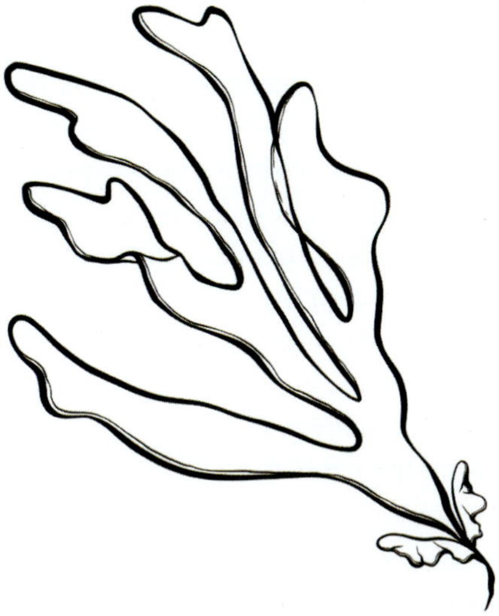

Palmaria mollis

Palmaria comes from the Latin word for "palm"; *mollis* is Latin for "soft."

KINGDOM: Plantae
PHYLUM: Rhodophyta
CLASS: Florideophyceae
ORDER: Palmariales
FAMILY: Palmariaceae

Atlantic dulse, *Palmaria palmata*, has been harvested and eaten dried as a snack in the Maritime provinces of Canada since the second half of the nineteenth century. Harvesting dulse in Nova Scotia is referred to as "dulsing," and a good low tide during dulse season is known as a "dulsing tide." Most people aren't aware that we have dulse on the Pacific coast as well. It is, however, a different species (*Palmaria mollis*), and the difference in texture and flavour between the two dulses is certainly detectable to a dulse connoisseur. Traditionally, people in the Maritime provinces who had weak constitutions and anemia were encouraged to eat the mineral-rich seaweed to help "build the blood." Dulse has also been traditionally used to treat herpes and to alleviate seasickness. In Ireland, where dulse is known as *duileasc* or *dillisk*, the rusty-red seaweed has been enjoyed as a snack and used as a medicine to rid one of intestinal worms for centuries.

DESCRIPTION

Palmaria mollis is a rusty red colour and slightly translucent. It has a very small holdfast and stipe and deeply divided blades that look like misshapen fingers. The blades can grow up to approximately forty centimetres in length. The female seaweeds are tiny, almost microscopic. It is theorized that this is to prevent inbreeding.

Pacific dulse drying on screens in my shop.

RANGE

Palmaria mollis grows in the low intertidal and subtidal zones from Alaska to central California.

NUTRITIONAL PROPERTIES

Dulse is one of the richest food sources of iron. It is high in vitamins A, B-complex, C and D, and is a rich source of minerals including iron, iodine, calcium, phosphorus, potassium, magnesium, manganese and other essential trace minerals. It is a good source of protein and fibre and an excellent source of essential fatty acids, including omega-3s.

AS A HEALING AID

Dulse contains potent antioxidants and helps protect the body from free-radical damage. It helps support the immune system and nourishes the adrenals. Dulse contains antibacterial and antiviral compounds, and studies have shown specific antiviral activity against herpes viruses, validating its traditional use to treat outbreaks (see Protection against Viruses, page 129). Anthelmintic properties have also been shown in studies (see Protection against Parasites, page 130), and indeed, dulse has been used traditionally to rid the body of intestinal worms. It helps calm the nervous system and is soothing to the lungs. It can help alleviate iron-deficient anemia and has been traditionally used as a remedy for seasickness.

HARVESTING

Pacific dulse, *Palmaria mollis*, grows in the low intertidal and subtidal zones. I harvest dulse at low tide in the spring and early summer. I leave several inches attached to the rock and then cut with my knife. I prefer to harvest dulse that is still covered by water so I can rinse as I go. By early summer, the low intertidal dulse is starting to look ratty from all the exposure to the air and the elements, but the subtidal dulse is still in great condition. As the tides get progressively lower following summer solstice, healthier patches of dulse can be reached. To dry, I spread the seaweed on screens, but dulse can be hung to dry as well.

CULINARY

People always ask me: "What is your favourite seaweed?" While I can say that kelp are my biggest passion, answering that question is just too hard. But if someone were to ask me what my favourite seaweed is to eat plain and dried, that's easy. It's dulse.

Pacific dulse has an intense, meaty flavour that imparts a powerful tingle of umami on the tip of the tongue. It has a chewy texture and a richness of iron that you can taste. It has been marketed as

Dulse is slightly translucent when held up to the sun.

"vegan bacon" for this very reason. There is no other seaweed that tastes anything like dulse. My favourite way to enjoy it is as a snack, dried and on its own, or with roasted nuts such as almonds. I will also add it as the "meat" in a sandwich. Chefs are enamoured of dulse's powerful flavour and have created delicious meals with it. Check the recipe section (page 184) for adventurous and flavourful dishes that use dulse.

Mermaid Envy

From the moment I learned about the existence of mermaids, I was intrigued. I became envious of these beautiful, wild and fiercely independent creatures—part woman and part fish—who could breathe under water and hence live in the sea.

I first learned about mermaids when a lushly illustrated version of Hans Christian Andersen's "The Little Mermaid" was read to me as a very young girl. Unlike Disney's story, this classic version of the fairy tale has a morbid twist: the mermaid does get legs so she can be with her true love, but for the rest of her life it will feel as though she is walking on razor blades.

I was convinced that I could breathe under water. I had a specific technique that I used to practise and refine in the bathtub. I still have a faint but lingering belief that my technique actually does work.

Later on, when Disney's film *The Little Mermaid* came out, I changed the lyrics of the song "Part of Your World," a song in which the mermaid Ariel sings of her deep longing to be on land, to describe my deep longing to live under the sea. I was convinced that I could breathe under water. I had a specific technique that I used to practise and refine in the bathtub. I still have a faint but lingering belief that my technique actually *does* work.

I used to dream, both awake and asleep, that if I dove into the black ocean at night, I would be able to breathe under water and live forever under the sea. There was something about the ocean at night that made it seem to have this particular magical quality.

One night, after an evening out in my university days, I stood on the historical Halifax pier and stared intently out at the water. It was perfectly black, the exact shade and texture that I had dreamed would enable me to breathe water, like a mermaid. Halifax Harbour is one of the deepest natural harbours in the world, with a depth of eighteen metres at low tide. I knew there was plenty of space for me to navigate my first breath of water and make my way out to the open sea. I stood up on the squared timber that lined the very edge of the pier. I pictured myself diving into the water, through the blackness to the other world that lay beneath. I imagined the cold water around me as I inhaled my first breath of water. And then . . . *relief*.

I could breathe water. I pumped my tail flukes, swimming out of the harbour and on toward my new home. A home where dolphins, whales, sea turtles, crabs, stingrays and perhaps even mermaids would become my new family.

CHRIS ADAIR PHOTO

Sea Spaghetti

***Gracilaria* spp. and *Gracilariopsis* spp.**
Gracilaria comes from the Latin word for "slender."

KINGDOM: Plantae
PHYLUM: Rhodophyta
CLASS: Florideophyceae
ORDER: Gracilariales
FAMILY: Gracilariaceae

Fondly known in the Pacific Northwest as "sea spaghetti," species of *Gracilaria* and *Gracilariopsis* can only be distinguished by expert phycologists. This pretty pink seaweed is known as *ogonori* or *ogo* in Japan, "Irish moss" in the Caribbean and *limu manauea* in Hawaii. Species of *Gracilaria* are cultivated and harvested to extract agar, which is used as a commercial thickener and stabilizer.

Sea spaghetti is a fleshy, translucent seaweed, which, depending on growing conditions, ranges in colour from greenish yellow to flamingo pink. In the cold, nutrient-rich waters where I live, sea spaghetti is always pink, making it easy to spot when it has been washed up onto the shore. It has cylindrical branches growing off a central cylindrical frond, which emerges from a small discoid holdfast. It is less than a metre in length.

This subtidal species quickly loses its vigour when exposed to air. EMMA GEIGER PHOTO

RANGE

Local species of *Gracilaria* and *Gracilariopsis* grow in the subtidal zone from northern BC to Baja California, Mexico.

NUTRITIONAL PROPERTIES

Sea spaghetti is a good source of vitamins A, B-complex and C, and minerals including calcium, magnesium, iron, sodium, phosphorus and potassium. Sea spaghetti is also a good source of protein, fibre and essential fatty acids.

AS A HEALING AID

Gracilaria species have antiviral and antibacterial properties and have shown specific antiviral activity against Japanese encephalitis virus. They are also a rich source of antioxidants.

HARVESTING

This is another seaweed I harvest exclusively by salvaging. It is a subtidal species that does not float, so to harvest it while still attached would require diving for it. It is a splendid treat to find fresh and vital sea spaghetti washed up on the shore. Since this seaweed has a very high percentage of water, it loses most of its form and substance when dried and is best consumed fresh. It will only last a day or two in the refrigerator unless you keep it suspended in clean seawater, which helps it stay fresh for three or four days.

CULINARY

Fresh sea spaghetti is a culinary delicacy. The texture, flavour and aroma are wonderful. Sea spaghetti has a texture reminiscent of cucumbers, crisp on the outside and juicy on the inside. Its mild, briny flavour is great as an addition to salads and rice noodles, and as a garnish on grilled fish.

A Sea of Pink

There is something about the colour pink that brings out the little girl in me. It is a colour that reminds me to love and care for myself with the fierce pampering of a grandmother.

Some days, when I walk out into the shallow waters of the intertidal zone at low tide, I am surrounded by a sea of pink. Rainbow seaweed, flashing her different colours like jewels. *Corallina*, her powder pink splashed upon the rocks she grows from. *Polyneura*, blushing pink with indigo iridescence running along her veins. Puce-hued *Callophyllis*, wine-red *Prionitis* and rusty-pink dulse. Delicate *Ptilota*, carnation pink and lacy, reminiscent of my great grandmother's crocheted doilies.

There is something about the colour pink that brings out the little girl in me. It is a colour that reminds me to love and care for myself with the fierce pampering of a grandmother. Pink hues capture my curiosity, my imagination and my childlike sense of wonder.

I feel uniquely nurtured, playful and cheerful on those sunny spring tides, wading in a shallow sea of pink.

Powder-pink coralline algae, iridescent rainbow seaweed, pale pink *Polyneura*/*Cryptopleura* and other rhodophytes add a symphony of colour to the intertidal zone on sunny spring days.

A blood star in shallow water blends in with diverse and bright-coloured rhodophytes growing in the low intertidal zone.

In the spring of 2018 I was harvesting *Alaria* with my apprentice Emma when she asked me what all the pretty blue shells were that she could see shimmering under the water. The only blue shells we have here on the coast are mussel shells, but I had never seen any in my *Alaria* garden before. I waded over to take a look and found that she was standing in a patch of rainbow seaweed that was shimmering and flashing beautiful hues of indigo. A smile stretched across my face.

"Those aren't shells," I said.

I proceeded to introduce her to rainbow seaweed, which I call "the magical seaweed" when I teach kids, because it truly is. After her first meeting with rainbow seaweed, Emma, an avid photographer, spent hours trying to capture its magic on film. There are indeed some beautiful photos of this seaweed, but none of them come close to the experience of actually seeing it in person. The beauty of *Mazzaella* is truly a *moving* beauty, one that can't be fully captured.

Rainbow Seaweed

Mazzaella's iridescence is more subtle out of the water.

Mazzaella splendens

Mazzaella comes from the name of the Italian phycologist Angelo Mazza; *splendens* is Latin for "shining, shimmering, glistening."

KINGDOM: Plantae
PHYLUM: Rhodophyta
CLASS: Florideophyceae
ORDER: Gigartinales
FAMILY: Gigartinaceae

DESCRIPTION

Mazzaella splendens is a seaweed of breathtaking, magical beauty. It is a deep, rich purple and is wonderfully iridescent. It shimmers mostly indigo, but if you see a young, healthy one in the water on a sunny day, you will see it shimmer not only indigo but also red, green and gold, like a trove of exotic jewels. It is a flat, bladed seaweed that is typically a plum purple, though in warmer, nutrient-poor waters it can be a yellowish green. It has a tiny holdfast from which a single, oval-shaped blade arises that reaches a maximum length of approximately seventy centimetres.

You may see the common name "rainbow kelp" in some seaweed books, but because it is not a kelp and not even a brown seaweed, I prefer "rainbow seaweed."

RANGE

Mazzaella splendens is found in the low intertidal to shallow subtidal zone from Alaska to Mexico.

NUTRITIONAL PROPERTIES

Mazzaella splendens is an abundant source of vitamins A, B, C and E and minerals including calcium, magnesium, iron, potassium, zinc, iodine and sodium. It is a good source of protein, fibre and essential fatty acids.

AS A HEALING AID

Mazzaella splendens applied topically or soaked with in the bath is nourishing and hydrating to the skin. It contains potent antibacterial, antiviral and antifungal compounds and is helpful for chronic

skin conditions, including psoriasis and eczema. It also helps with skin irritations, including rashes, acne, and dry, itchy or irritated skin. It can be applied to burns to soothe, prevent infection and promote healing. The essence of rainbow seaweed is used to treat depression.

HARVESTING

I harvest most of my *Mazzaella splendens* by salvaging. I often find beautiful pieces washed up when I am heading out to my *Alaria* garden. I do harvest the odd piece that is still attached, though. I feel particularly aware, when harvesting *Mazzaella*, of a sacredness, and only feel comfortable cutting a small amount each season. I personally leave about half of the seaweed attached to the rock and am very selective about which ones I harvest. I then dry the seaweed on screens, though you can also dry by hanging on a rope or in tree branches. Once dry, I store rainbow seaweed in mason jars that I keep by my tub.

Mazzaella splendens has the silkiest texture of any seaweed I have ever touched. When I first began harvesting, I sometimes used texture to positively identify seaweed species. If the rainbow seaweed is young and lacking iridescence, beginners could easily mix it up with dulse, since both occupy the low intertidal to subtidal zones. However, if you touch the two specimens, dulse is rougher and rainbow seaweed feels like silk. If your rainbow seaweed feels rougher and lacks the silky texture, it is past its prime.

CULINARY

There is very little information available on the edibility of *Mazzaella splendens*. It has a bitter flavour compared with other local seaweeds and has immense therapeutic value when bathed with and used topically. For these reasons, I reserve my rainbow seaweed for use in the bath.

SPA

Rainbow seaweed is absolutely wonderful to bathe with and deeply therapeutic. In warm, fresh water, its surface dissolves into a silky spreadable gel, which I rub onto my skin, especially on my face and around my eyes. You can leave it on after your bath is done to receive even more benefit.

Seaweed Apprentices

Knowledge of the craft of sustainably harvesting wild seaweeds in the Pacific Northwest is very rare, and when I wanted to learn, there was no one to teach me. I felt a responsibility to teach those who had a passion to learn, which is why I decided to take on apprentices. As a private introvert, and one who treasures the solitude that can only be experienced harvesting seaweed alone, it was a daunting venture. However, my own experience—coupled with a desire to spread sustainable and ethical harvesting practices—led me to teach others the craft of wild seaweed harvesting.

I teach people of all ages and diverse backgrounds about seaweeds in a variety of settings, including public schools, colleges and universities, community halls, conference centres and private spaces. With apprentices, I get to really dive deep in imparting knowledge. While the actual harvest is hands-on learning without much talking, the hours of hanging seaweed in the seaweed shop allow for lots of questions, contemplation and seaweed philosophy.

There is something remarkable about spending day in and day out with someone who shares your passion and is eager to learn. I have many cherished memories of watching a student enter the kelp forest for the first time, or touch their first piece of *Alaria*. The unique connections I have made with my seaweed apprentices are among the highlights of my career. My apprentices teach me the freshness of new wonder. They ask tremendous questions that challenge me to look at things in new ways and motivate me to seek even further.

My apprentices teach me the freshness of new wonder. They ask tremendous questions that challenge me to look at things in new ways and motivate me to seek even further.

Apprentice Emma Geiger harvesting *Alaria marginata*, winged kelp.

"In spending time with seaweed and harvesting with Amanda, handling the long golden slippery fronds of kelp, I felt drawn in by the intimate qualities of seaweed . . . I was so gently taken into the world of these mysterious beauties from the sea."

—LIZ GLOWACKI, BULL KELP APPRENTICE

"I can recall how I evolved from liking the harvesting to loving it. It is as if the *Alaria* is trying to seduce you with her elegant silken blade, dancing to the current of the ocean. Being in the ocean, surrounded by all the seaweeds, is like melting in with them, like there is nothing else in that moment other than the seaweeds, who are winking at you. Pure mindfulness.

Not only did Amanda show me her love for seaweed but I found her passion for it infectious. Through her I internalized a more natural lifestyle by foraging and making my own products. The most important lesson I learned was do what you love doing and own it."

—KELLY VAN HOUTTE, *ALARIA* APPRENTICE

"Wild harvesting bull kelp with Amanda entails gearing up in a hooded wetsuit with booties, gloves, snorkel, and getting into the frigid Pacific Ocean. The first time we went into the water to harvest bull kelp, Amanda was very cautious, watching the water's movements and making sure I could handle it. Once I was comfortable, she took me out to the kelp forest. I screamed so much into my snorkel, I choked a few times. The phenomenal beauty of each golden green (think green chartreuse) blade of bull kelp floating majestically under water is a sight to behold. I have never seen anything more magnificent in my life. Every single time I saw the kelp forest, my heart would burst with immense joy. It felt like the kelp forest was radiating a magical love energy."

—TIEN BUI, BULL KELP APPRENTICE

Tien and me after the bull kelp harvest. SALLY SHEN PHOTO

Bleach Weed

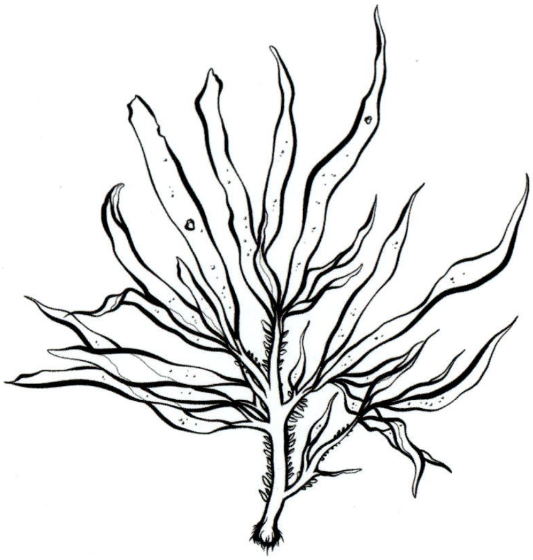

EMMA GEIGER PHOTO

***Prionitis sternbergii* (formerly *Prionitis lyallii*)**
Prionitis is derived from Greek and means "like the holly" or "saw-edged."

KINGDOM: Plantae
PHYLUM: Rhodophyta
CLASS: Florideophyceae
ORDER: Halymeniales
FAMILY: Halymeniaceae

DESCRIPTION

I love to find this seaweed when I am giving seaweed tours. I have participants smell it and tell me what they think it resembles. Bleach weed, like the name implies, releases a scent when rubbed that smells like chlorine bleach. Indeed, it contains significant amounts of chlorine, believed by phycologists to be a deterrent for grazing herbivores.

Prionitis sternbergii is a deep wine-red colour and has flattened blades branching off from a middle rib, though in warmer, nutrient-poor waters its colour can be a yellowish green. This perennial seaweed grows up to forty centimetres long.

RANGE

Prionitis sternbergii grows in the mid to low intertidal zone and is often found in tide pools. It grows in wave-exposed and semi-protected shores from Alaska to Baja California, Mexico.

AS A HEALING AID

Species of *Prionitis* have been studied for their antiviral properties.

CULINARY

Bleach weed is not considered an edible species.

MAY 2017

The Cabezon

It was a warm, sunny day in late spring, and I was out harvesting *Alaria*. The tide was low, and the outcropping of very large boulders that I harvest near was completely dry. I was rounding the tip of the outcropping when a noise startled me.

I whipped around to see what it was and looked into the eyes of a river otter, who looked equally startled, just a few feet away.

The otter ran a few steps to the tide line and then quickly disappeared under water. I looked back to the rock where it had been standing, which was surrounded by tall boulders on three sides. There, on a seaweed-covered rock, lay a large cabezon that the otter had been eating when we surprised each other.

The huge, bulbous head was still intact and the fish's eyes, which appeared too big for its already large head, were wide open. The body and tail, which appeared too small and skinny to belong to the head, were half eaten.

I carefully, with my wetsuit-booted foot, nudged the fish, wanting to roll it over and see the other side.

I screamed in alarm as the fish, miraculously still alive, clamped its jaws and sharp teeth down on my foot. I frantically shook my foot around and the fish came off.

Thank God for 8mm booties.

"How could this poor fish still be alive?" I wondered incredulously. It was on dry land and clearly had been for some time now, and its body was half eaten. "Nature can indeed be cruel," I thought.

The most humane thing to do for this fish would be to kill it. But when I thought about having to smash it with a rock, my whole being vibrated with fear and panic. I'm the person who finds a bug or spider in my house and carefully puts it outside, offering it well wishes in its new home. I can't kill things.

"It's a fish out of water," I rationalized. "It will die momentarily."

I placed a large piece of kelp over it to protect it from anything coming to eat it again and to allow it to die without the stimulation of its bright, sunlit surroundings. Its huge eyes remained wide open. To me, they expressed the acute fear and horror of the fish's experience and, to this day, the sight of those eyes is still etched into my memory.

I screamed in alarm as the fish, miraculously still alive, clamped its jaws and large teeth down on my foot.

I left the cabezon to hopefully die quickly and more peacefully. I proceeded to harvest *Alaria*, choosing smooth, beautiful pieces from under the waist-deep water, scrubbing along the midrib under water with my bare hand and cutting carefully several feet above the sporophylls.

When my bag was full, I made my way over the slippery seaweed to my large harvest bag with backpack straps, emptied my smaller bag into it, and headed back out to the low intertidal and subtidal zones to fill another bag. En route, I stopped to check on the cabezon.

EIKO JONES PHOTO

It was still alive! "How can that be possible?" I wondered. It had been at least twenty-five minutes since I had first found it. I felt sick with horror that this poor fish was still in agony.

After I filled my next bag, the ill-fated fish was finally out of its misery. I felt so relieved.

As I drove home from the beach that day, I recalled the experience. My mind had a myriad of thoughts racing through it: the wonder of seeing a cabezon so close up, the horror when I realized that the fish was being eaten alive and the internal conflict I felt about not having the courage to put it out of its misery. It was hard to let myself imagine the pain and fear of the fish's last moments. As I pulled into my driveway I thought, "What a bizarre 'office' I have." I never know what kind of experience I will have at work.

As I hung each harvested piece of *Alaria* on cedar racks in my drying shop, my head continued to be full of thoughts about that cabezon.

Turkish Washcloth

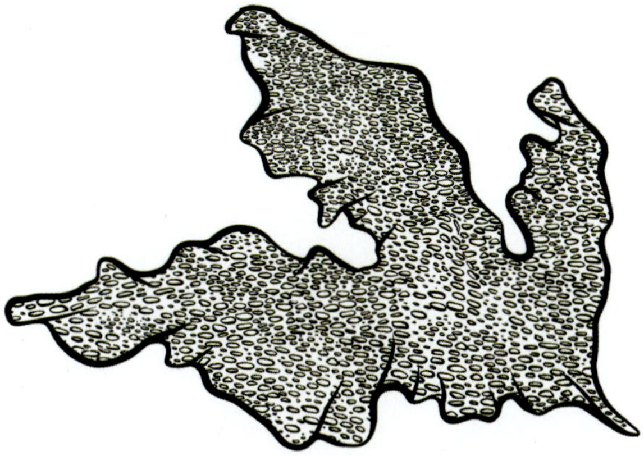

***Mastocarpus* spp.**

Mastocarpus comes from the Greek word for "whip-like body."

KINGDOM: Plantae
PHYLUM: Rhodophyta
CLASS: Florideophyceae
ORDER: Gigartinales
FAMILY: Phyllophoraceae

DESCRIPTION

Ten species of *Mastocarpus* are reported to be found in the Pacific Northwest. *Mastocarpus* species are all small, very dark purple seaweeds that usually have a blade with a single but subtle division that is shaped like a Y. Blades are typically less than fifteen centimetres long, curled and covered with papillae, though they lack papillae when very young. *Mastocarpus* species have a crustose phase in their life cycle.

RANGE

Mastocarpus species grow in the mid to high intertidal zone from Alaska to California.

AS A HEALING AID

Mastocarpus species, rich in mucilaginous carrageenan, are healing and soothing to the trachea and lungs. They contain potent antiviral, antibacterial and antifungal compounds and can aid in healing, soothing and preventing infection in tissues that are inflamed due to bronchitis, pneumonia, damage from smoking and other respiratory ailments. In Ireland, *Mastocarpus stellatus* is used interchangeably with *Chondrus crispus* (known in many regions as "Irish moss") for medicinal purposes. Traditionally, it was taken with warm milk as a treatment for respiratory ailments and consumed to ward off colds and flus.

HARVESTING

I harvest *Mastocarpus* species at low tide in the late spring and summer. Where I live, on the west coast of Vancouver Island, they aren't really large enough to start harvesting until the second half of June. Even at their largest, they are small seaweeds. Also, the older and larger each seaweed is, the more defined the papillae are, and since I use them as an exfoliant and skin tonic in the bath, that is what I am looking for. I also harvest *Mastocarpus* by salvaging. I find beautiful pieces washed up in the fall and early winter when I walk the beaches with my dog. *Mastocarpus* species are very easy to dry. I dry them on screens in my drying shop during my harvest season, but in the winter, when my shop is shut down, I just spread them out on the counter and they dry in a day or two.

The bumps on *Mastocarpus* species make them perfect for use as a "scrubby."

The crustose phase of *Mastocarpus* spp. can be seen on the left of the photo and looks like a dark purplish-black splotch on the rock. The erect phase is on the right.

CULINARY

Mastocarpus species are considered edible, but their culinary use is primarily as a thickener due to their rich carrageenan content. When soaked and cut, they will form a thick gel.

SPA

Like *Chondracanthus* species, *Mastocarpus* species are valued for their use in thalassotherapy (bathing in seaweeds). The common name, Turkish washcloth, reflects this, and *Mastocarpus* species share the morphological feature of papillae with the larger Turkish towel. They make an excellent natural exfoliant that can gently scrub away dead skin cells. I keep a Mason jar filled with dried *Mastocarpus* beside my tub and use it as a natural "scrubby."

Mastocarpus can be used several times in the bath if hung to dry between uses.

Nature's Masterpiece

"They play and make jokes and love one another . . ."

—ERNEST HEMINGWAY

The ocean is sacred to me. Each drop of it is holy water. When each of my two daughters were born, I took them down to the beach and rubbed drops of sacred, salty ocean into their foreheads: that was how I "baptized" them. When I surf, I say a prayer to the ocean each time I enter and again when I leave. Indeed, your fate is in her hands each time you leave the shore.

I believe my love of the ocean is intrinsic, that I was born with it. It is part of my soul, stretching back many eons before I was born.

Every act of passion I have made as a steward of the ocean can be traced back to my childhood love of dolphins. That love is the source of a life lived, as much as possible, in the sea.

But I didn't dream of growing up to be a seaweed harvester. It was dolphins who had my heart as a child. Every act of passion I have made as a steward of the ocean can be traced back to my childhood love of dolphins. That love is the source of a life lived, as much as possible, in the sea.

I grew up in one of the most landlocked places on earth: southern Ontario. I had a rubber dolphin when I was very young, and before I was even of school age, I had a love of dolphins that was so much deeper than any rational understanding. Totally pure.

This love began before I had ever actually laid my eyes upon a real dolphin. Their beauty and perfection in every way make me overflow with emotion that Mother Earth could have created such a thing. Smooth, velvety skin and warm, chocolatey eyes. Their bodies a shimmery and alluring shade of grey that you could get lost in, tapering to the most delicate of pinks on their bellies. A wonderfully powerful tail fluke that, unlike a fish, whose tail moves side to side, pumps up and down. A champion breath-hold diver and an air-breathing mammal that nurses and nurtures its babies, and to whom family is everything. A brilliant swimmer with a smiling face, impeccable sight and hearing and the magical superpower of echolocation.

And a natural love of and curiosity about humans. I read books documenting real-life stories of deep human-dolphin friendships in the wild, lasting years and, in some cases, a lifetime. I read true stories of dolphins fiercely protecting humans from sharks and rescuing victims of shipwrecks. My heart didn't know which way to break when I learned that the US Navy enlisted dolphins to work on submerged explosives, capitalizing on their magical sonar and incredible diving abilities.

When I was eighteen, I spent a week assisting the Dolphin Child Project, in which a child psychologist used dolphin touch in their work with severely disabled children. I watched a child progress, both developmentally and physically, as if by magic; this child achieved goals in one week of working with the dolphins that he hadn't been able to achieve in years of trying.

This took place at the Dolphin Research Center in the Florida Keys. The dolphins there lived in the ocean, with only underwater plastic fences to keep them in, which served more to keep other creatures out. One time a shark got in, and by the time the staff were able to get it out, the shark was fully traumatized. Marina and Aleta, the female dolphins whose pen the shark had ended up in, had a wonderful time playing with their new "toy."

The dolphins at the centre were mostly retired aquarium dolphins. When dolphins get too old to perform, they are often sent to pools "behind the curtain" and largely forgotten about. Oftentimes, they are isolated in these pools. Dolphins are highly social creatures, and it is painful to imagine the anguish they must experience being isolated, not only from their own kind, but from any ocean creatures.

Dolphins use echolocation, a natural form of sonar, to "see" their surroundings, to locate fish in the wild and possibly to communicate. Although we know that dolphins communicate with vocal sounds such as clicks and whistles, some cetologists think that echolocation clicks are also used by dolphins to express their emotions. When dolphins use echolocation to "see" or locate, they send out clicks and then wait and listen to the echo, which then sends a message to the brain, essentially painting them a picture of what their clicks have touched. Their echolocation is so precise that they can determine minor differences in density. I witnessed one dolphin, a lovely female named Molly, differentiate a dime from a quarter. In a concrete tank, dolphins' echolocation clicks reverberate endlessly, rendering one of their main forms of sensing their world unusable.

My passion for seaweeds was preceded by a passion for the ocean that can be traced back to this childhood love of dolphins.

When I was a little girl, I wanted to be a dolphin trainer when I grew up. My first trip to MarineLand in Niagara Falls fixed that. Seeing these beautiful animals, for whom I had the purest kind of love and saw as *nature's masterpiece*, in concrete pools when the most beautiful place on earth, the ocean, was their rightful home, was too much for my innocent heart to bear. Watching them do stupid tricks for food was like rubbing salt in the wound.

I spent the whole visit bawling my eyes out, staring with a broken heart into their sad yet smiling eyes. The visit did not unfold the way my parents had predicted.

I proclaimed to the dolphins that day that I would make it my life's mission to free every cetacean from captivity, and that I would not stop until every last one was free. If and when I accomplished that, I would move on to freeing all other animals from cages. After all, freedom is a natural and universal need and right.

You gotta love a child's idealism.

As I got older, I stayed true to that mission. While I was in university, working toward my bachelor of science in marine biology, I formed a group with several other like-minded folks called the Animal Liberation Collective. Though I had been a vegetarian since I was nine and was aware of some of the crowded and unpleasant conditions farm animals endure, it was with this group that I really had my eyes opened to the horrific treatment of caged animals on industrial farms. The scale and brutality of what factory farm animals endure was clearly the worst animal abuse on the planet, and it quickly became the main focus of our group's work. In addition to the animal suffering, I learned that raising animals on an industrial scale requires and contaminates massive quantities of fresh water, is the largest driving force of tropical deforestation and is even one of the largest contributors to climate change. Unfortunately, global meat consumption has increased more than fourfold in the second half of the twentieth century and by 2050 it is predicted to be twice the amount it was in 2008, between 460 and 570 million tons.

My despair began to be replaced by hope after meeting Mandy Carter in 2018 at a forum called "The End of Cages" at Terra Madre Salone del Gusto, the biannual Slow Food gathering in Italy. Mandy worked for an organization called Compassion in World Farming, which had launched a European Citizens' Initiative calling upon the European Commission to propose legislation to ban the use of cages in animal farming, and she related that it was receiving overwhelming support.

In Canada we have enacted a total ban of dolphins and whales in captivity. On June 10, 2019, Canada's House of Commons passed the "Free Willy" bill. Otherwise known as Bill S-203 or the Ending the Captivity of Whales and Dolphins Act, the bill prohibits the wild capture, captivity and breeding of whales, dolphins and porpoises, and restricts trade in these animals as well.

We have a long way to go as a society, but there are some positive changes happening. The people who dedicate themselves and work so hard toward positive change are all heroes of mine.

My intrinsic, childhood love of dolphins sent me on a compassionate journey of discovery and instilled in me a deep sense of personal responsibility to protect what I love. When I teach, I encourage people to discover what they love, and then share their passion with others. I believe sharing your passion is one of the most effective ways to protect what you love. People are rarely inspired to act—to change their habits—because of statistics, no matter how scary they are. People are inspired to act when they feel joy and an emotional connection to something.

Following your passion, your curious longing, your intrinsic love, can lead you to places you would never have dreamed. For example, a childhood love of dolphins could lead one to create a life among the mystical kelp forests and seaweeds of the Pacific Northwest.

Turkish Towel

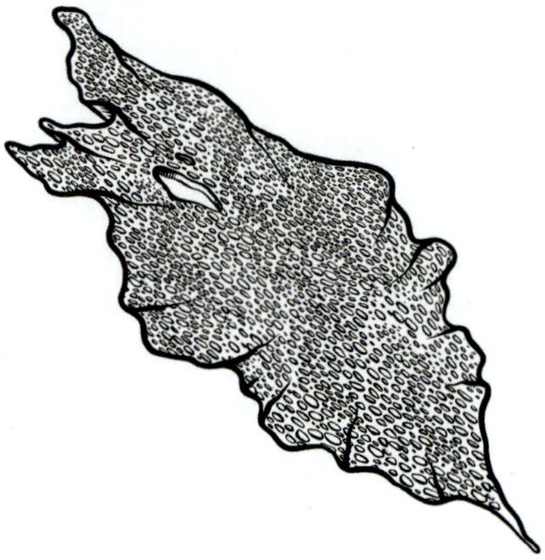

Chondracanthus spp.

Chondracanthus is derived from Greek words meaning "spiny cartilage."

KINGDOM: Plantae
PHYLUM: Rhodophyta
CLASS: Florideophyceae
ORDER: Gigartinales
FAMILY: Gigartinaceae

DESCRIPTION

Turkish towel is a beautiful, rich purple seaweed (though in some warmer, nutrient-poor waters it can be a greenish yellow). It has a small, discoid holdfast and a very short stipe, and it grows into a single blade reaching lengths of approximately eighty centimetres. It is covered with papillae that are not believed to be associated with reproduction, but are just part of the seaweed's regular morphology. The two outer sides of the blade can be pulled apart, revealing a fleshy, gelatinous interior. The most common species of *Chondracanthus* in the Pacific Northwest are *C. exasperatus* and *C. corymbiferus*.

RANGE

Chondracanthus species grow in the low intertidal and subtidal zones from Alaska to Mexico on the west coast of North America.

AS A HEALING AID

The rich carrageenan of *Chondracanthus* species has been used traditionally to treat chronic respiratory ailments such as pneumonia, bronchitis, emphysema and damage from smoking. *Chondracanthus* species also contain potent antiviral compounds, and studies have reported antiviral activity against HIV and HPV (see Protection against Viruses, page 129).

HARVESTING

I have never seen Turkish towel attached and growing on the beaches where I live on the west coast of Vancouver Island, though I have in other areas of the island. Hence, I do all of my harvesting of this treasured seaweed by salvaging. Every year, in November, December and January, I find beautiful

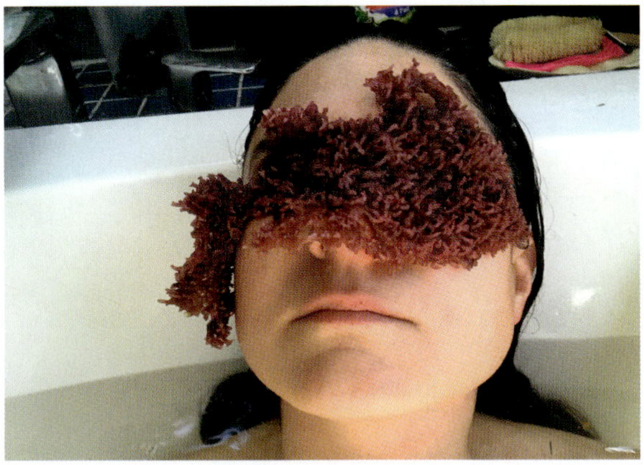

Turkish towel hydrates and tones the skin.

pieces of Turkish towel washed up on my home beaches. The winter storms rip up fresh, healthy and vital pieces, and I gather them, put them in my pockets and just lay them out to dry on my counters when I get home.

CULINARY

Though *Chondracanthus exasperatus* is considered edible, its culinary use is primarily as a thickener due to its rich carrageenan content. When soaked and cut, it will form a thick gel.

SPA

Chondracanthus species are such valued seaweeds for thalassotherapy (bathing in seaweeds) and spa use that they were named for it: Turkish towel. The bumps covering the seaweed's exterior make it a natural exfoliant that can gently scrub away dead skin cells. When exposed to fresh water in the bath, its rich carrageenan begins to ooze out. This gel can be applied to the face and body to improve skin tone, fade blemishes, hydrate and nourish the skin, and provide a measure of UV protection.

Two beautiful pieces of Turkish towel harvested by salvaging.

AUGUST 2018

Disco Jellyfish

It was the year of the jellyfish. That bull kelp season, I saw more jellyfish, by both numbers and diversity, than any other season: tiny cross jellyfish (*Mitrocoma cellularia*) would suddenly appear beside my mask; dozens of moon jellies (*Aurelia labiata*), glowing with an ethereal whiteness, drifted among the bull kelp, pulsing like hearts in slow motion; stringy scalloped jellies (*Solmissus* spp.) floated at different depths in the water column like a collection of strings; and fiery red lion's mane (*Cyanea capillata*), with their star-shaped medusas, arrived in great numbers with the still heat of August. Toward the end of August, mid-harvest on a very hot day, I saw a jellyfish that looked just like an egg frying in a frying pan! My fine observational skills were confirmed when I arrived home, looked in my field guide to see what species it was and found that the common name for *Phacellophora camtschatica* is "fried egg jellyfish." The most stunningly beautiful jellyfish I saw that season, one I had admired in books but had never seen in the ocean before, was the sea nettle (*Chrysaora fuscescens*). It had pale pink tentacles—shaped like soft corals and hanging far beneath its amber medusa—and these were encircled by wispy, string-like ones. It pulsed with an alluring beauty that found me gently undulating along behind it, my task at hand momentarily forgotten.

There was one jellyfish, though, that made me question my own mind. I first saw it on a bright, sunny day, toward the beginning of the bull kelp season. I was snorkelling out to the kelp forest when I spotted it. I may not have noticed it at all as it was quite small, but its internal light show caught my peripheral view. It was flashing

It was flashing a vibrant red, blue, green and yellow light internally, one colour at a time, from top to bottom, in quick succession. I stared at it, squinting my eyes inside my mask, trying to make some sense of what I was seeing.

a vibrant red, blue, green and yellow light internally, one colour at a time, from top to bottom, in quick succession. I stared at it, squinting my eyes inside my mask, trying to make some sense of what I was seeing. What I was witnessing did not fit in with what I knew about living beings. I stared in disbelief, expecting the illusion to disappear, yet as I stared, it continued flashing the bright colours in rapid succession. It made me think of disco lights.

When I arrived home, I hit the books, not really expecting to find confirmation of what I *knew* I had seen. Lo and behold, I discovered that the lobed sea gooseberry (*Bolinopsis infundibulum*), a type of comb jelly, can do precisely what I had seen. It has synchronized cilia attached to bioluminescent plates that account for the disco-esque light show it produces.

Few creatures appear more likely to have come from another planet than jellyfish. With their transparent bodies, the simultaneous movement of so many distinct, delicate parts makes me think they are plugged into some otherworldly rhythm.

Other Common Red Seaweeds of the Pacific Northwest

AHNFELTIOPSIS SPP. >
Ahnfeltiopsis species look like little deer antlers and are often found partially buried in the sand.

< CALLOPHYLLIS SPP.
Callophyllis species could be confused with *Palmaria mollis*, Pacific dulse, by a novice seaweed investigator.

CORALLINA VANCOUVERIENSIS >
Corallina vancouveriensis, a coralline algae, looks like a pretty pink fern. EMMA GEIGER PHOTO

< CRYPTOPLEURA OR HYMENENA SPP.
It's hard to tell whether this seaweed is a *Cryptopleura* or *Hymenena* species. These two genera are difficult to distinguish in the field, even for a seasoned phycologist. In *Pacific Seaweeds*, Druehl and Clarkston note that these two genera are easily confused and "in need of taxonomic investigation." When held up to the sunlight, many faint veins can be seen within the blades. EMMA GEIGER PHOTO

HALOSACCION GLANDIFORME >
The common name "dead man's fingers" is not an inaccurate description of this bizarre-looking seaweed, which is filled with seawater. Tiny holes allow the water to be sprayed from its "fingers" when squeezed, making it an effective short-range squirt gun. EMMA GEIGER PHOTO

< *HILDENBRANDIA* SPP. WITH *PSEUDOLITHOPHYLLUM MURICATUM*
This *Hildenbrandia* species looks like a blood-red stain on the rock. The coralline algae *Pseudolithophyllum muricatum* also appears in this photo and looks like powder-pink splotches.

JOHANSENIA MACMILLANII >
When you see the coralline algae *Johansenia macmillanii* growing, it is easy to see why biologists previously thought it was a coral.

< *MICROCLADIA COULTERI*
The common name "sea lace" alludes to the delicate beauty of *Microcladia* species.

< *NEMALION VERMICULARE*

I have only seen this species once, on a trip to Bamfield, a remote community on Vancouver Island that is home to the distinguished Bamfield Marine Sciences Centre. It is reminiscent of purple earthworms, but without the rings. Indeed, its species name, *vermiculare*, comes from the Latin word for "worm."

NEORHODOMELA LARIX >

Neorhodomela larix has small branches growing radially on its main branches, much like the foliage of a larch tree, and hence acquired the common name "black larch." Its other common name is "black pine."

< *ODONTHALIA FLOCCOSA*

Odonthalia floccosa is a delicate, highly branched red seaweed.

OSMUNDEA SPECTABILIS >

Osmundea spectabilis has a gelatinous sheen.

EMMA GEIGER PHOTO

PTILOTA SERRATA >

Ptilota serrata is another beautiful, delicate red seaweed.

< *SPARLINGIA PERTUSA*

Sparlingia pertusa, pinkish purple and covered with tiny, delicate holes, is one of the most beautiful seaweeds on our coast. EMMA GEIGER PHOTO

Order Corallinales—Coralline Algae

Coralline algae contain calcium carbonate in their cell walls, which makes them rigid. They play an essential role in the ecology of coral reefs and are an important part of the reef's structure. They used to be classified as corals, which are animals, until it was discovered in 1837 that they are red algae. They are very pretty seaweeds, most often a delightful shade of powdery pink.

CHRIS ADAIR PHOTO

PHAEOPHYCEAE

THE BROWN SEAWEEDS

"Let your spine be loose like seaweed."

—EOIN FINN

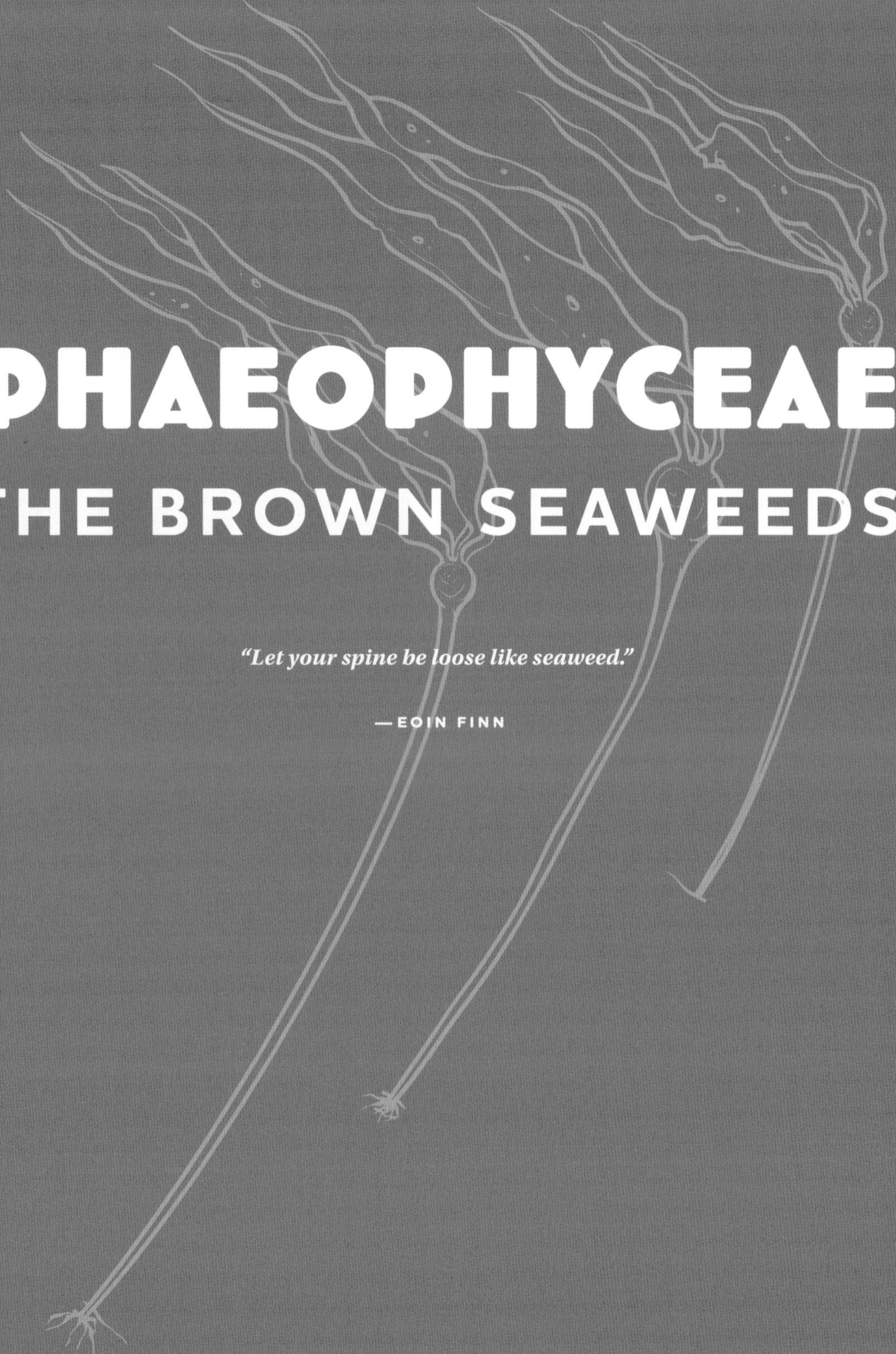

Bull Kelp KELP

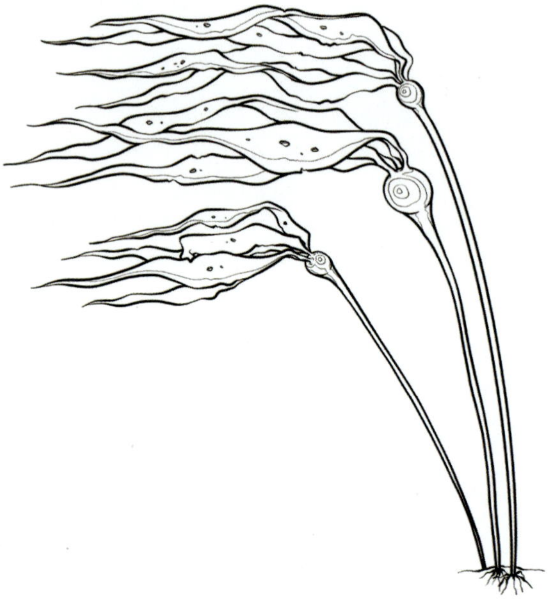

Nereocystis luetkeana

Nereocystis is derived from Greek words for "mermaid" and "bladder"; *luetkeana* is named for the German-Russian explorer and scientist Captain Friedrich Benjamin von Lütke.

KINGDOM: Chromista
PHYLUM: Ochrophyta
CLASS: Phaeophyceae
ORDER: Laminariales
FAMILY: Laminariaceae

DESCRIPTION

Bull kelp, an iconic seaweed of the Pacific Northwest, is the second-largest seaweed in the world. It has a holdfast made of many haptera (finger-like projections that look like roots) from which a hollow stipe emerges, starting off narrow and becoming progressively wider until it culminates in a gas-filled bulbous float called a pneumatocyst. Growing off the top of the pneumatocyst are the fronds. The first several inches are tightly branched and the same colour as the stipe, and then they transition into long flat olive-green blades. The stipe can reach lengths of thirty-six metres and the blades can reach lengths of up to four metres. Incredibly, as an annual, bull kelp achieves its impressive size in a single growing season of about six months, making it one of the fastest-growing organisms on the planet! It is able to grow thirty centimetres per day on sunny days in summer.

On some of the blades there are reproductive spore patches that appear as dark, rectangular shapes that are thicker than the surrounding blade. The youngest spore patch is closest to the stipe end of the blade and as you progress toward the distal end of the blade, where the patches are more mature, they become larger, thicker and darker. Each spore patch contains millions of spores, and once fully mature, the patch is released, leaving a rectangular hole in the blade. Quite amazingly, the spores are equipped with little tails called flagella, which enable them to swim the rest of the way to the ocean floor. Close to the holdfast of their parent seaweed, the spores then develop into male and female gametophytes and with the union of male and female sex cells, fertilization occurs. The resulting zygote develops into a new sporophyte, which is the macroscopic phase of their life cycle and what we recognize as a bull kelp.

Bull kelp, along with giant kelp, form the canopy and primary structure of the Pacific Northwest's kelp forests, world renowned for their biodiversity, ecological productivity and beauty.

Some of the blades of bull kelp contain thick, rectangular spore patches (sori).

RANGE

Bull kelp only grows in the northeastern Pacific Basin, from Alaska to Point Conception, California.

NUTRITIONAL PROPERTIES

Bull kelp is a rich source of vitamins A, B-complex, C, D, E and K and minerals including calcium, magnesium, iodine, potassium, bromine, phosphorus, chromium, zinc and iron. Bull kelp is an excellent source of both soluble and insoluble fibre and essential fatty acids, and it is approximately one-quarter protein.

AS A HEALING AID

Studies have shown that the rich sodium alginate content of bull kelp helps to safely remove certain heavy metals, radiation, dioxins, polychlorinated biphenyls (PCBs) and other toxins from the body (see Toxin Cleansing, page 128). Bull kelp contains fucoidan, fucoxanthin, phlorotannins and other biologically active compounds that studies suggest help to prevent certain types of cancers and chronic diseases such as cardiovascular disease (CVD) and type 2 diabetes (see Seaweeds and Chronic Diseases, page 134). It contains potent anti-inflammatory compounds and can be useful in relieving inflammation associated with arthritis and other conditions, both by consuming it and by using it topically. Bull kelp supports healthy thyroid function, aids digestion and has antiviral activity, with specific activity against the herpes family of viruses. Bull kelp can help reduce symptoms associated with mental illnesses such as attention-deficit/hyperactivity disorder (ADHD) and is abundant in nutrients essential to healthy brain function.

NOTE: Members of the kelp family, including bull kelp, should only be used *following* radiation treatment, not during, due to their effectiveness at removing radiation from the body. The nutritional benefits of seaweed are provided by consuming a small amount regularly, about five grams per day. If you plan to use bull kelp or other seaweeds therapeutically, be sure to consult with a medical professional first.

HARVESTING

Bull kelp is a subtidal species, meaning it is always covered by water, even at the lowest tides. It requires cold, moving water to thrive. Young bull kelps will not survive if they are anchored in water that is too shallow, allowing them to be regularly exposed to the air. Several hundred metres offshore from the beach where I live, healthy bull kelp forests grow each year, becoming tall enough to reach the surface in late June. To harvest bull kelp, I suit up in a 5mm wetsuit with hood, booties and gloves. I wear a mask, fins and snorkel and take a large mesh dive bag with shoulder straps and a pocketknife, and I swim out to the forests. If the bull kelp is large enough, the gas-filled bulbous float and blades float at the surface. The blades grow from the bulb in four main sections, and I only harvest from two of those sections, leaving fifteen to thirty centimetres of blade still attached to the bulb. If the stipe is cut, the seaweed will not regenerate. After cutting, I "swoosh and stuff" the blades into my harvest bag, gathering all the floating ends and pushing them in. As much as possible, I try to harvest blades that do not have spore patches.

Harvesting bull kelp. CHRIS ADAIR PHOTO

When my bag is full, after harvesting from approximately five individual seaweeds, I swim back toward the shore, take off my fins in shallow water, strap the bag onto my back and wade onto the beach. There, I deposit my bag of seaweed into my wheelbarrow and head back out for another load.

In the fall, most of the spore patches have been released. At this time, there is not much left of many of the blades, and they quickly become susceptible to grazing herbivores. This signifies the end of my harvest season, which for bull kelp runs from July to September. The stipe, being much thicker and more robust than the delicate blades,

Gathering the cut blades into my harvest bag. CHRIS ADAIR PHOTO

Transporting the bull kelp harvest back to my shop for hanging. AGATHE BERNARD PHOTO

Making kelp pickles.

persists much longer. While surfing in the winter, I sometimes see the floating stipes with no blades, and they make me think of mermaids who have lost all of their hair.

The stipe of bull kelp is edible, but it does not dry well like the blades do, and therefore must be consumed fresh or preserved as pickles. Because bull kelp will not regenerate if the stipe is cut, I

Tuna tataki created by chef Manpreet Sethi at Aura restaurant, located inside the Inn at Laurel Point in Victoria. The dish is made with Dakini Tidal Wilds' bull kelp, cured albacore tuna, avocado mousse, compressed cucumbers, local cherry tomatoes and yuzu fluid gel.

MANPREET SETHI PHOTO

harvest stipes by salvaging. A day with a big wind or a big swell in July will typically pull up some stipes in perfect condition and wash them right onto the beach. Two medium-sized stipes are enough to make about eight litres of pickles.

CULINARY

In my opinion, bull kelp tastes the saltiest of all the edible seaweeds and has a very strong umami flavour, which feels like a delightful tingling on the tip of the tongue. This extraordinary flavour, combined with a light texture that melts in the mouth, makes bull kelp an excellent seasoning, flavour enhancer and base in sauces. I make a delicious pesto with bull kelp by grinding it up in a blender and then adding pepper, garlic, extra-virgin olive oil, freshly squeezed lemon and sometimes pecans and shaved parmesan cheese. I keep a jar of bull kelp flakes beside my stove and use it like I would salt on top of most meals. Be aware that spore patches taste different, particularly mature ones that are much thicker than the rest of the blade, and they are generally not preferred.

The Gift

I am a seaweed harvester. I spend the summer months snorkelling in the icy cold waters of the Pacific Northwest's kelp forests. I live almost as far west as you can get in Canada, on the west coast of Vancouver Island, and I see the kelp forests out of the windows of my home.

In my thirteenth season, while snorkelling in a kelp forest harvesting bull kelp, I was given a rare gift in the form of an experience with another creature.

I suited up in my 5mm wetsuit and headed down to the beach with my wheelbarrow, knife, harvest bags, mask, fins and snorkel. It was a hot summer day, and I felt the usual thrill of anticipation as I prepared to slip into the enchanting underworld of the Pacific Northwest's kelp forests. I walked backward with my fins into the Pacific, turning around in chest-deep water to rinse and pull on my mask. I melted into the sea while the sounds of the outside world retreated until there was only the sound of my breathing through my snorkel.

sometimes, after a harvest, when I blink. However, the mystical enchantment of being immersed in them makes each time feel like my first.

On days when there is no current, even if it is quite windy at the surface, the hair of the mermaids—the blades—falls straight down toward the seaweed-covered bottom. The ground beneath the kelp forest is thickly covered by *Cymathaere triplicata*, *Alaria marginata*, *Pterygophora californica* and powder-pink splotches of coralline algae. On these days, it is somewhat possible to weave by surface into the kelp forest.

Most often, though, there is a good current cruising east from the west. This causes the mermaids' hair to weave an impenetrable tapestry of blades. On these days, diving beneath this golden-green canopy is the only way to get in or out of the forest. This day was such a day. I was surrounded by thick, healthy bull kelp on all sides, taking in the wonderment before I started to fill my bag.

Peering through my mask, I saw the head of a giant Pacific octopus resting on the thick blades of twenty-five-foot-long bull kelp.

I watched the beige, rippled sand, with the odd Dungeness crab doing its sideways walk, turn to swaths of red laver, *Alaria*, *Cymathaere* and dulse, until finally the bull kelps, with their round bulb heads and hair-like blades, dancing like mermaids in the current, filled my entire field of vision.

I had arrived. The bull kelp forests are so familiar to me that I see them in my dreams, and

And then, I saw it. Peering through my mask, I saw the head of a giant Pacific octopus resting on the thick blades of twenty-five-foot-long bull kelps.

Its head was the size of my pillow and a deep wine red. Its eye was closed, as though sleeping, and the apparatus just below the eye was opening and closing rhythmically with a cadence much like the sound of my own breath through my snorkel.

Its head lay just a few feet from my face. I truly couldn't believe what I was seeing. A sense of deep gratitude washed over me. I knew this was a very special gift.

I allowed my gaze to move beyond the head of the octopus. Like something out of a movie, I saw tentacles from this mystical giant poking in and out of the bull kelp blades as far as I could see. Lines of large, white suction cups disappeared into the thick tangle of blades, only to reappear in another location. Shades of red ringed the white cups, which suctioned onto the lime green of the bull kelp. It didn't feel real.

I lay there suspended in the water, staring in wonder, my head just a few feet away from this wondrous, otherworldly creature. I have no idea how much time passed. It was exactly like they say when you are completely consumed with an experience: time stood still.

Eventually I remember thinking, "Well, I guess I'd better get to work." The awesome humour of that statement was definitely not lost on me.

I pulled myself away, looking back often until the blades of the bull kelp that surrounded me were free of tentacles, despite my imagination struggling to let them go. Once I had filled my harvest bag, I kicked back to the shore, the kelp forest disappearing behind me, then the thick seaweed carpet giving way to sand and eventually to shore. In chest-deep water I took off my fins, pulled up my mask and gingerly walked out of the sea with my heavy load. A whole other world lay behind me.

I dropped my harvest bag on the beach rocks, let some of the water drain out, then humped it up to my wheelbarrow and headed back in for my second bag.

I weaved beyond the edge of the forest and once again found myself staring at the enormous, wine-red head of the sleeping octopus. I was over the moon.

This experience illuminated a place in my heart I didn't know was there. It forged an emotional connection with a creature based not on common ground or natural instinct, but on an overwhelming sense of wonderment and respect, instilled in me because of my experience with it.

The octopus I saw that day, a giant Pacific octopus (*Enteroctopus dofleini*), is the largest species of octopus in the world. I learned that although octopuses are molluscs they are incredibly intelligent. They are able to learn quickly and use tools and, one of my favourite indicators of intelligence, they enjoy a diverse diet and have individual preferences. I also learned that the tentacles I thought I had seen were, in fact, arms.

The camouflaging ability of octopuses, the ability to change both their colour and texture in an instant to match their surroundings, is the stuff of science fiction. Although I consider myself something of an escape artist, they are truly the masters, able to squeeze through a keyhole one suction cup at a time. In fact, shortly after my experience, an octopus named Inky escaped from his aquarium in New Zealand, crawled across the floor, squeezed himself into a drainpipe and found freedom back in the ocean!

Perhaps closest to my heart, I learned of the fierce protection a mama octopus gives her babies before they are at the mercy of the big blue. At the end of a female octopus's life, she finds a safe haven and lays many rows of eggs. As her body slowly breaks down, not unlike a spawning salmon, she stands guard over her unborn young until she has almost no vitality left. Once her young have hatched and joined the realm of zooplankton, she dies within moments, having hung on just long enough to send her young off on their own oceanic journey.

A couple of years after my experience in the kelp forest with the giant Pacific octopus, I got a tattoo. It was during a challenging time in my life, and my experience with the octopus kept surfacing. That moment was a great gift, and although it was seemingly random, it would not have happened if it weren't for years of dedication. My tattoo of the octopus weaving with the bull kelp serves as a permanent reminder to trust myself and my path, and the unexpected gifts that can come as a result.

CHRIS ADAIR PHOTO

The kelp forests of the Pacific Northwest are a global treasure. They have an eerie and otherworldly beauty and support an unquantifiable plethora of life. They are the underwater forests of the temperate rainforest region of the emerald-hued Pacific Northwest. CHRIS ADAIR PHOTO

Kelp Forests

"The number of living creatures of all Orders, whose existence intimately depends on the kelp, is wonderful. A great volume might be written, describing the inhabitants of one of these beds of sea-weed . . . I can only compare these great aquatic forests . . . with the terrestrial ones in the intertropical regions. Yet if in any country a forest was destroyed, I do not believe nearly so many species of animals would perish as would here, from the destruction of the kelp."

—CHARLES DARWIN, *THE VOYAGE OF THE BEAGLE*

Kelp forests are an oasis of life. They are one of the most productive ecosystems in the world and support an astounding level of biodiversity. Here in the Pacific Northwest, we have more species of kelp than anywhere else in the world, approximately thirty species out of an estimated hundred worldwide. While most temperate coastal waters have kelp forest ecosystems, in most regions they would be better described as kelp shrubs. The largest underwater "trees" in the world form the canopy of the ethereal, mystical and grand underwater forests of the Pacific Northwest. These two special species of kelp are the largest species of seaweed on earth: *Macrocystis pyrifera* (giant kelp) and *Nereocystis luetkeana* (bull kelp). They are incredibly fast growing and both are capable of reaching lengths of forty metres, and even longer for giant kelp. Large kelp provide structure and are primary producers of these temperate oceanic forests. The kelp forest is home to a wonderful spectrum of life, including thousands of species across multiple phyla. In the Pacific Northwest, creatures such as abalones, urchins, sponges, crabs, univalves, bryozoans, jellyfish, fish, seals, sea lions, seabirds, orcas and so many more live in this thriving ecosystem.

Kelp require cold water and a rocky bottom in order to anchor and grow. Certain animal species also play an important role in determining the presence or absence of kelp forests. For example, in the Pacific Northwest the main grazers of kelp are purple sea urchins (*Strongylocentrotus purpuratus*). Urchins are such voracious kelp eaters they can clear-cut a kelp forest at a rate of nine metres per month, according to NOAA. What is left behind is almost bare rock. This forlorn-looking seascape is called an urchin barren. Because I am fortunate to live and harvest among robust, healthy bull kelp forests, I had never seen an urchin barren until I travelled to the Bamfield Marine Sciences Centre to teach a class. Upon taking the students snorkelling in a kelp forest, I saw one for the first time. It was even more dramatic than the photos I had seen. The transition from forest to urchin barren was so abrupt and distinctive you could have laid down a rope to divide the two ecosystems. On one side, dense, canopy-forming bull kelps averaging six metres in length thrived. On the other side, there were silt-covered rocks and what looked like close to one hundred urchins just in my field of vision. I was astonished.

Urchin populations are often a direct indicator of the presence or absence of kelp, and the balance between urchins and their predators is critical to the health of kelp forests. Sea otters, sunflower stars and wolf eels are the main predators of urchins in the Pacific Northwest. Sea otters in particular can be such efficient eaters of urchins that they are considered a keystone species. In areas where urchins exist, the presence or absence of sea otters can directly determine the presence or absence of kelp forest ecosystems. Sea otters were extirpated from the Pacific Northwest at the turn of the twentieth century due to an insatiable appetite for their fur, which is the densest of any fur-bearing animal. Sea otter fur was sold primarily to China, parts of Europe, Russia and the US. Sea otters are currently making a significant comeback in some areas, which is having a positive influence on kelp forest ecosystems. Dr. Lynn Lee, a biologist studying the impact of sea otters on kelp forests in Haida Gwaii, has noted that it is primarily "rafts of males," groups of young male sea otters who have recently left their mothers and are fending for themselves for the first time, who eat urchins because they are easy prey. According to Dr. Lee's research, females tend to eat harder-to-catch and preferred prey, such as crabs and shellfish.

Kelp forests and kelp beds have among the highest rates of primary production of any ecosystem in the world and grow along 25 per cent of the world's coastlines. They are currently in a global state of decline at an estimated rate of 2 per cent annually (Wernberg et al., 2019). At the Phycological Society of America's 2018 meeting at the University of British Columbia in Vancouver, a number of kelp phycologists presented research suggesting that warming ocean temperatures are the most significant driver of the global decline of kelp forests.

Kelp can only grow in cold water. Braeden Schiltroth, a young phycologist at Simon Fraser University, studies the effect of temperature on bull kelp in the Salish Sea, off the southwest coast of British Columbia. Schiltroth's research shows that sea surface temperatures (SSTs) of 17C have a negative effect on the bull kelp's spore formation, and at 20C, spores are killed. A temperature difference of only *three degrees* is the difference between bull kelp surviving or perishing. These temperature increases are already occurring in the waters of the Strait of Georgia, where thick bull kelp forests that existed into the late 1970s and early 1980s have mostly disappeared. A few areas of the Strait of Georgia still have healthy bull kelp forests, usually in areas with strong currents, but most of the bull kelp that remains is sickly and small. Although a number of factors have led to the massive decline in bull kelp forests in the Strait of Georgia, some of which are still unknown, studies suggest that increased SSTs play a major role.

The kelp forests of the Pacific Northwest are breathtaking and possess a mystical beauty and vitality. CHRIS ADAIR PHOTO

Extreme weather events such as marine heatwaves (MHWs) are another significant global stressor that can have a catastrophic effect on kelp forest ecosystems. In contrast to the relatively slow warming of the world's oceans, an MHW occurs when a specific area of the ocean experiences abnormally warm temperatures over a relatively short period of time. The longest-lasting and largest MHW occurred in the northeast Pacific Basin. It covered a massive swath of the northeastern Pacific, extending from Alaska to Baja California, Mexico. With some estimates in excess of four million square kilometres, its size dwarfed all other MHWs on record. Known as "the blob," it was first detected in 2013, persisted to the end of 2015, and appeared to dissipate in 2016. The blob caused unusual weather on the west coast of North America and disrupted all coastal ocean ecosystems in its range, affecting every level of the food chain. Populations of fish, marine mammals, seabirds and other marine life were devastated due to the blob's nutrient-poor water and the arrival of and subsequent competition from new warm-water species. The warmer water temperatures also created the perfect conditions for toxic algae blooms, which became more prevalent. Alarmingly, an MHW similar to the blob appeared in the same area of the north-eastern Pacific Basin in 2019.

In addition to the direct negative effect of the blob's warmer temperatures, sea star wasting disease (SSWD) is also devastating to kelp forests. The disease affects many species of sea stars and other echinoderms, causing them to slowly die by essentially disintegrating. Despite efforts by scientists, very little is known about SSWD, but it is believed that the blob played a significant role in its prevalence in 2013 and 2014 by providing a warm-water environment more advantageous to pathogens. The massive die-off of sea stars at that time has been described as "one of largest marine die-offs in modern history" (Iovenko, 2019). Along with other species of sea stars in the Pacific Northwest, sunflower stars were extirpated. Because sunflower stars are important predators of urchins, their absence helped lead to an explosion of urchin populations that further ravaged kelp forests. According to a study published in the journal *Scientific Reports* in 2019, scientists from the University of California, Davis, and the California Department of Fish and Wildlife discovered that there was a sixty-fold increase in urchin populations between 2014 and 2015. They also found that since the arrival of the blob in 2013, more than 90 per cent of bull kelp forests along 349 kilometres of coastline in Northern California have been wiped out (Kerlin, 2019).

At the 2018 meeting of the Phycological Society of America, I learned about another particularly deadly MHW and its effects on kelp forest ecosystems while listening to a presentation by phycologist Thomas Wernberg from the University of Western Australia. This MHW occurred in 2011 off the coast of Western Australia where dramatic increases in SSTs wreaked havoc on ocean ecosystems and killed huge swaths of kelp forests and kelp beds hundreds of kilometres long. These kelp-based ecosystems still have not recovered, and instead have transitioned from kelp forest ecosystems to turf ecosystems. Turfs lack structure-providing kelp, which are replaced with low-lying seaweeds. A turf ecosystem does not have the structure or primary productivity (the speed at which solar energy is converted to organic compounds) needed to support the biodiversity of a kelp forest. Unfortunately, regime shifts from kelp forests to turfs are becoming more prevalent in many regions as ocean temperatures continue to rise and the occurrence of MHWs increases.

In addition to the negative effect of global stressors such as increased ocean temperatures, local stressors such as pollution, storms, coastal development, invasive species, harvesting and fishing can also have a detrimental effect on kelp forest ecosystems, often compounding the severity of global stressors. Where I harvest, a major threat to the health of kelp forests is the possibility of an oil spill. In what is unfolding as Canada's version of the Standing Rock protest, Kinder Morgan, an oil company based in Texas, proposed the building of a 980-kilometre-long pipeline that would twin an existing pipeline. If built, the pipeline would run from the tar sands in northern Alberta to a marine terminal in Burnaby, just outside Vancouver. From there, super tankers would ship the diluted bitumen to Asian markets. If the project goes ahead, the Juan de Fuca Strait, where I harvest, would see a seven-fold increase

Sunflower star in winged kelp. Sunflower stars (*Pycnopodia helianthoides*), the largest species of sea star in the world, were extirpated from the Pacific Northwest in 2015 due to sea star wasting disease. Since sunflower stars are a predator of urchins, their absence has had an effect on kelp forests.

in the number of oil tankers navigating the rugged body of water. This project, called the Trans Mountain Expansion Project (TMX), is among the most divisive political issues in Canada I have witnessed in my adult life. The provincial government of BC, as well as several BC municipalities, including Vancouver, BC's largest city, and Victoria, the provincial capital, are opposed to the pipeline, specifically to its related increase in oil tanker traffic along BC's coast. First Nations and environmental non-profit groups have organized many protests blockading several areas along the pipeline's route, one of which saw the arrest of two politicians from two different parties.

But in a stunning move, the federal government of Canada bought the project in 2018 with $4.5 billion of taxpayers' money when Kinder Morgan threatened to pull out due to several ongoing court challenges opposing the project. To date, eight federal court challenges have been filed by the municipalities of Vancouver and Burnaby, the Tsleil-Waututh, Squamish, Kwantlen and Coldwater First Nations and the provincial government of BC. The six First Nations won their case in the Federal Court of Appeal in 2018: the courts agreed that they were not adequately consulted before the pipeline was approved.

If the TMX proceeds, it will support the development of the Alberta tar sands. There are an estimated 170 billion barrels of oil in Alberta's tar sands that can be extracted with modern technology, and an additional 1.63 trillion barrels could be extracted if it were possible to separate all the available bitumen from the sand. James Hansen, former climatologist at NASA, describes the tar sands as "one of the dirtiest, most carbon-intensive fuels on the planet" (Biello, 2013).

In the midst of all this, young phycologist Ranah Chavoshi from Simon Fraser University conducted a study on the effects of diluted bitumen (the oil product that would be travelling through the pipeline and on the supertankers) on the developmental stages of the brown seaweed *Fucus distichus*. Diluted bitumen is considered more toxic than oil and often contains benzene, a known human carcinogen. Not surprisingly, Chavoshi's results showed that diluted bitumen had a significant negative effect on the growth of the brown seaweed. I feel deeply inspired and hopeful to see young phycologists like Chavoshi doing such timely and relevant work.

Another local threat to kelp, and indeed other seaweeds in the Pacific Northwest, is clearcut logging. The mountainous coastal ecosystem of the Pacific Northwest is temperate rainforest, which is among the rarest and most endangered ecosystems on the planet. As a true rainforest, it

Surface view of a bull kelp forest on a foggy day.

receives a high amount of seasonal precipitation, which frequently leads to landslides. The roots of trees provide stability to slopes, essentially holding everything in place. When all of the trees are removed, such as in areas that have been clearcut, the occurrence of landslides increases greatly, especially on steep slopes. Landslides throw large amounts of sediment down creeks and rivers to the ocean, where it can smother seaweeds, blocking their ability to receive light and photosynthesize. Heavy sediment can also cover rocks, which seaweeds need in order to attach their holdfast and grow. Moreover, the covering of cobblestones with sediment in creeks and rivers also renders salmon breeding grounds unsuitable for egg laying, further disturbing the delicate balance of the coastal ecosystem of the Pacific Northwest.

Although the global decline of kelp forests is disheartening, the good news is that kelp is starting to get some real recognition for its critical role in the health of ocean ecosystems and of the world at large. My hope is to inspire a personal connection to these environmental stewards, and, in turn, create a sense of personal responsibility for their protection. As Jacques-Yves Cousteau noted, "people protect what they love."

The Colour of God

The experience of being in a kelp forest is a spiritual one. Every time I am out there feels like the first time. Every time I am overwhelmed with awe and wonder. I feel I am in the presence of the Great Spirit . . . of God.

I become consumed in this white-green glowing light. I know it is healing depths that are far beyond my conscious awareness.

CHRIS ADAIR PHOTO

The ice-green, glacial colour of the water appears to have a white glow from within. I become consumed in this white-green, glowing light. I know it is healing depths that are far beyond my conscious awareness. Growing out of this otherworldly and spiritual light are the bull kelps themselves. With the sun and the water filtering dancing rays upon their blades, their ancient, eerie beauty is a thing to behold. They dance and sway to the rhythm of the swell and the currents—the rhythm of the sea.

Each time I arrive at the edge of the glowing green underwater forest, I find myself in disbelief that a thing of such mysterious beauty even exists.

I am so grateful.

When I am among the kelp forests, immersed in the green, internally glowing light, I know without hesitation that life is a gift. It is an adventure meant to be explored with an open heart.

"I must be a mermaid . . . I have no fear of depths and a great fear of shallow living."

—ANAÏS NIN

CHRIS ADAIR PHOTO

Giant Kelp KELP

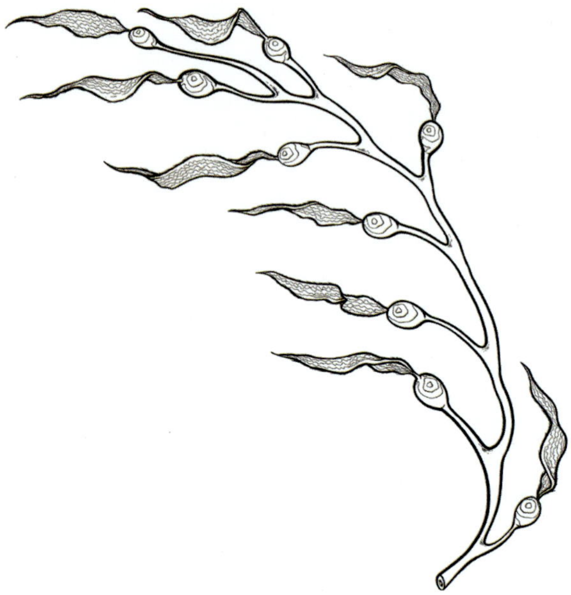

Macrocystis pyrifera
Macrocystis is derived from the Greek words for "large" and "bladder."

KINGDOM: Chromista
PHYLUM: Ochrophyta
CLASS: Phaeophyceae
ORDER: Laminariales
FAMILY: Laminariaceae

DESCRIPTION

The aptly named giant kelp is indeed the largest seaweed on earth. Alternating float-bearing corrugated greenish-brown blades grow from a central stipe. The branches closest to the holdfast lack floats and are reproductive. The genus *Macrocystis* originally contained three species, but genetic analysis has resulted in the current recognition of just a single species, *M. pyrifera*. What used to be considered two different species in our waters, *M. pyrifera* and *M. integrifolia*, are now considered different ecotypes (genetically distinct populations within a species) with differences in morphologies and habitats. *Macrocystis pyrifera* is a perennial seaweed, but due to the severity of winter storms, it rarely persists longer than three years. This giant seaweed is reputed to reach lengths of fifty-five metres.

Giant kelp is one of the two canopy-forming species that create the Pacific Northwest's beloved kelp forests. Indeed, stands of this great seaweed look like giant underwater trees.

RANGE

Macrocystis pyrifera grows primarily in the subtidal zone from Alaska to Mexico, though in some areas it can persist into the low intertidal zone.

Harvesting salvaged *Macrocystis* in winter. JENNIFER JELLETT PHOTO

NUTRITIONAL PROPERTIES

Giant kelp is a rich source of vitamins A, B-complex, C, D, E and K and minerals including calcium, magnesium, iodine, potassium, bromine, phosphorus, chromium, zinc and iron. Giant kelp is an excellent source of both soluble and insoluble fibre, protein and essential fatty acids.

AS A HEALING AID

Giant kelp is a nutritional powerhouse and rich in antioxidants. Like all kelp, it contains sodium alginate, which safely cleanses toxins and radiation from the body. It is a source of fucoidan and fucoxanthin, which research suggests increase immune function and help to prevent chronic illnesses such as cancer, diabetes and CVD (see Seaweed and Chronic Diseases, page 134). Its richness in iron and other supplementary minerals make it useful for treating anemia, and its iodine content supports healthy thyroid function.

HARVESTING

Where I live, giant kelp grows quite far offshore, so snorkelling for it is not a good option. Therefore, this is another seaweed I harvest by salvaging. Big bundles of beautiful, untorn and healthy-looking giant kelps wash up on very windy days or when there is significant swell. If you are harvesting from

Vera Rønningen's kelp chocolates made with *Macrocystis*.

seaweeds that are still attached, be sure to leave the reproductive blades closest to the holdfast and several feet of stipe and blades attached.

CULINARY

Giant kelp is a wonderful edible seaweed with a light texture and a mild, briny flavour. It pairs beautifully cooked with whitefish such as halibut and lingcod. My friend Vera Rønningen makes chocolates with *Macrocystis* that are breathtaking both in flavour and aesthetics. If you're lucky, you may find her selling them at the farmer's market on Hornby Island, BC.

SPA

Giant kelp is lovely to use in the bath and is very therapeutic and rejuvenating. It calms and soothes the central nervous system, aids in the elimination of toxins and lactic acid and improves circulation and skin tone (see Seaweed and the Skin, page 150).

My nephew gifted me this dried bull kelp stipe he found at the beach. He called it an "alien time signature." Algae are the oldest existing lifeforms on earth and it is theorized that all life evolved from algal ancestors. Forests of kelp existed in the ocean long before there were forests on land. Perhaps seaweeds are a key to a deeper and more ancient source of connection.

Walking Kelp KELP

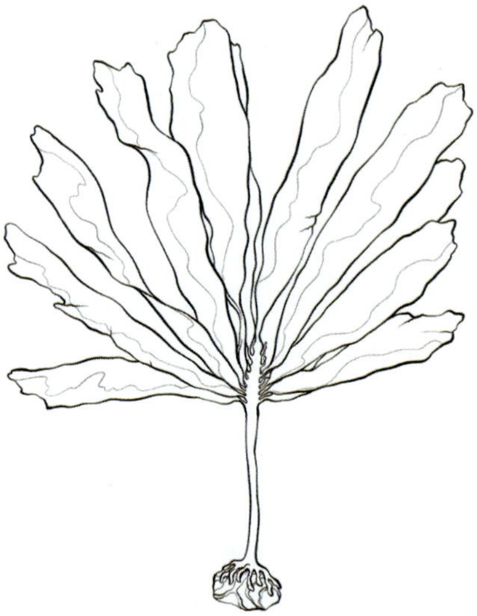

Pterygophora californica

Pterygophora is derived from the Greek word for "having wings."

KINGDOM: Chromista
PHYLUM: Ochrophyta
CLASS: Phaeophyceae
ORDER: Laminariales
FAMILY: Alariaceae

DESCRIPTION

Walking kelp is one of the longest-lived perennial seaweeds, surviving for up to twenty-five years. It is considered the old growth in kelp forests and, like trees, it can be aged by counting rings inside the woody stipe. The common name of this seaweed describes its rare ability to move locations. Walking kelp anchors to rocks with a diameter of about ten to fifteen centimetres. Surges caused by large swells can move the rocks that *Pterygophora* is anchored to, giving the appearance of the kelp having walked. Their haptera-style holdfast is so tightly attached to the rock that one can pick up the seaweed by the woody stipe, rock and all. In some coastal First Nations, walking kelp was used

The stipe of *Pterygophora californica* appears like wood when seen lying on the beach, but if you pick it up, it is much more flexible. EMMA GEIGER PHOTO

EMMA GEIGER PHOTO

to create channels at the mouths of rivers, to trap salmon returning to spawn.

Pterygophora californica can grow up to two metres long. The blades, which grow laterally from the top portion of the stipe, are reproductive sporophylls. At the terminal end of the stipe is a single vegetative blade with a subtle midrib.

RANGE

Pterygophora californica grows in the subtidal zones of exposed and semi-exposed beaches from Alaska to Mexico.

HARVESTING

I don't harvest walking kelp because it is a long-lived perennial and makes up the old growth of the kelp forests, playing a critical role in the longevity of the ecosystem. Additionally, the blades are reproductive sporophylls and it is not considered an edible seaweed.

On the beach where I live, subtidal walking kelp gets dragged closer and closer to shore with the huge swells and windstorms of winter, bringing with them the rocks that their holdfasts are still tightly gripped onto. As the summer solstice approaches and the tides become lower and lower, they begin to find themselves on dry land. The rocks they are attached to are often so deeply buried in the sand that they can't be budged.

Orca Baby

My youngest daughter was born on July 17, in the middle of the seaweed harvest season. I harvested *Alaria* that year until the end of June. For about the last three weeks of that *Alaria* season, while I was eight months pregnant, I watched a large pod of orcas fishing and swimming as I harvested. Most likely, it was the Southern Resident Killer Whales (SRKW) I spent time with that summer in Orveas Bay. To this day, I have never seen orcas so consistently, for so long. Though they were always pretty far out, it looked to be a pod of about fifteen to eighteen individuals. My baby would be entering the world for the first time in the coming weeks, and I was among orcas every day, fully immersed in their energy.

The orcas of the Pacific Northwest truly look at home when viewed against a backdrop of misty, temperate rainforest–covered mountains in chlorophyll-green water.

Orcas are the largest member of the dolphin family and the top predator in the ocean. Though they hold this mighty title, there are no reports of an orca ever harming a human in the wild. The only instances of a human being harmed and even, in a couple of cases, killed by an orca, have happened in captivity. I don't even want to consider the psychological hell of being ripped from my family and home and made to live out my days in a concrete cage.

With their dramatic black-and-white colouring, orcas are recognized the world over. The orcas of the Pacific Northwest truly look at home when viewed against a backdrop of misty, temperate rainforest–covered mountains in chlorophyll-green water.

In British Columbia, we have two distinct types of orcas: the residents and the transients. When I was in university in the late 1990s, it was theorized that the residents were in the process of speciation, meaning that they were transitioning to a distinct species. There are two populations of resident killer whales: the northern residents, formally known as the Northern Resident Killer Whales (NRKW) and the SRKW. Whereas transients eat marine mammals, primarily seals and sea lions, resident orcas are exclusively fish eaters. The critically endangered SRKW eat a diet of mainly Chinook salmon. It is estimated by cetologists that their diet is made up of 90 per cent Chinook salmon, primarily from the Fraser River population. Hence, the health of this salmon population and the health of the SRKW are intimately and critically connected. At the time of writing in 2021, there were seventy-five individuals left in this distinct population and the lowest returns of Chinook salmon in recorded history.

Alexandra Morton, a marine biologist and orca expert, is on a quest to try to save the SRKWs. She has followed this quest to the plight of the salmon,

My youngest daughter Nesika, fourteen years later, in the bull kelp forest I named the "Nesika Garden." AGATHE BERNARD PHOTO

upon whom the resident orcas depend for their survival. What she has discovered is that deadly viruses and sea lice from open-net pen salmon farms are spreading disease to wild salmon—and many of those salmon farms are in the migratory path of the Fraser River Chinook salmon. She had these powerful words to say about the plight of the SRKW: "If we lose the Southern Residents, it will be the first extinction where every individual's name was known" (The Marine Detective, 2019).

Before orcas evolved to have their iconic black-and-white colouring, they were grey, like most other whales. A small area of this grey remains behind the dorsal fin. This grey area is called a "saddle patch" and is unique to each orca. Saddle patches are used to identify individuals. Every member of the SRKW and the NRKW has had their saddle patch photographed and recorded.

Unlike transient orcas, resident orcas live in large matriarchal societies. The grandmother is the wise leader of each pod that makes up a population. She holds knowledge of the language and is the navigator who knows the best fishing grounds. Matriarchs can live for many years after they are no longer fertile. The health and survival of the pod is dependent upon the matriarch and her knowledge.

Female resident orcas stay with their maternal family for their entire life.

"... saw an eagle soaring over the creek, saw a seal sunbathing on a rock, saw whales frolicking in the ocean, and a woman about eight months pregnant, wearing hip waders, out in the waves. And I was waiting when she waded ashore."

To harvest *Alaria*, I wear a wetsuit and wade in water between knee- and chest-deep, depending on the tide. As my season began that year, in early May, I started out wearing my regular wetsuit. A few weeks later, I was wearing a larger men's wetsuit, and by the end of the season, I was wearing my scuba dry suit, pulled up only to below the belly.

Paul Vasey, former host of "On The Island," a CBC radio show about Vancouver Island and the Gulf Islands, was on the beach where I was harvesting early one morning, working on another story. His curiosity piqued, he approached me as I was gliding carefully over the seaweed-covered rocks with my full harvest bag, heading for shore. He asked if I would mind being interviewed, and when I answered "Sure," he proceeded to ask me what on earth I was doing. His story begins like this: "... saw an eagle soaring over the creek, saw a seal sunbathing on a rock, saw whales frolicking in the ocean, and a woman about eight months pregnant, wearing hip waders, out in the waves. And I was waiting when she waded ashore."

I wrapped up my *Alaria* harvest season and three weeks later, my youngest daughter was born. About two weeks after that, I was back out, harvesting bull kelp. Though normally I would go to the very large forest down the beach, that summer I harvested in a smaller forest, close to the trail. I called it the Nesika Garden, named after my youngest daughter, because I could be out and back within an hour and a half to nurse her.

Both of my daughters have spent countless hours on the beach, going to work with Mom. I have listened to squeals of delight when they discover a new creature, and I have met creatures new to me, introduced by my daughters. I don't feel there is any better playground for kids than the beach: pure, salty air, playful waves and infinite things to discover underneath each rock.

Winged Kelp KELP

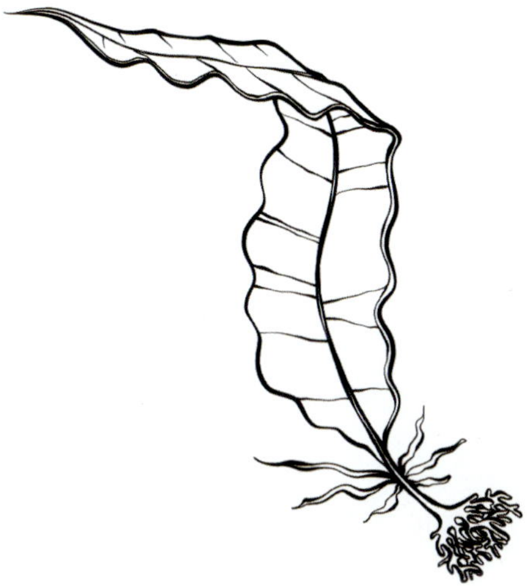

Alaria marginata
Alaria is Latin for "winged."

KINGDOM: Chromista
PHYLUM: Ochrophyta
CLASS: Phaeophyceae
ORDER: Laminariales
FAMILY: Alariaceae

Holding an *Alaria marginata* blade in my left hand and a sporophyll in my right. EMMA GEIGER PHOTO

Winged kelp with its holdfast attached to the rock, the sporophylls spread laterally from the stipe and the blade's thick midrib visible. EMMA GEIGER PHOTO

DESCRIPTION

Alaria is a beautiful biennial seaweed that has a branched holdfast, a short, cylindrical stipe and a long single blade with a well-defined midrib. It can grow up to approximately four metres in length. Reproductive sporophylls grow laterally from the stipe. They lack a midrib and are thicker than the vegetative blade. *Alaria* is similar, both biologically and nutritionally, to Japanese wakame (*Undaria pinnatifida*), as they both belong to the same family, and it is sometimes called "wild wakame."

RANGE

Alaria grows in the low intertidal and subtidal zones from Alaska to Northern California.

NUTRITIONAL PROPERTIES

Alaria is a rich source of vitamins A, B-complex (including B12), C, D, E and K. It is one of the highest food sources of calcium and is also rich in magnesium, phosphorus and potassium, as well as trace minerals including manganese, chromium, zinc, iodine and iron. *Alaria* is approximately one-quarter protein and contains essential fatty acids including omega-3s in an ideal one-to-one ratio. It is rich in sodium alginate and fucoxanthin and is one of the highest known sources of fucoidan.

AS A HEALING AID

Alaria helps to detoxify the body of certain heavy metals, radioactive compounds, dioxins, PCBs and other toxins via its rich sodium alginate content. It has been used traditionally to tone and cleanse the cardiovascular system, a use current research supports (see Seaweed and Cardiovascular Disease, page 137). *Alaria* nourishes and balances the female endocrine system and can be useful for relieving unwanted symptoms of premenstrual syndrome and menopause. Adding *Alaria* to soups and broths, especially miso broth, can aid in relieving the side effects of chemotherapy. When used following radiation treatment, it helps to restore essential nutrients and vitality. *Alaria* contains compounds that research suggests inhibit tumour growth and exhibit anticancer activity (see Seaweed and Cancer, page 134). *Alaria* is a rich source of antioxidants and anti-inflammatory agents and promotes healthy thyroid function. It is a traditional remedy used to prevent and ease colds and flus, with current research validating its antibacterial and antiviral properties (see Protection against Pathogenic Bacteria, page 129, and Protection against Viruses, page 129).

NOTE: Members of the kelp family, including winged kelp, should be used *following* radiation treatment, not during, due to their effectiveness at removing radiation from the body. The nutritional benefits of seaweed are provided by consuming a small amount regularly, about five grams per day. If you plan to use *Alaria* or other seaweeds therapeutically, be sure to consult with a medical professional first.

Fresh *Alaria* hanging to dry. EMMA GEIGER PHOTO

HARVESTING

Alaria, a biennial seaweed that lives in the low intertidal and subtidal zones, is best harvested in the spring and very early summer. Once the tides start getting lower as summer solstice approaches, *Alaria* is exposed to air more often. This can result in it being scorched by the sun and more susceptible to grazing by herbivores. In order to harvest sustainably, it is critical to leave the sporophylls intact and to cut the seaweed blade several inches above the stipe. As *Alaria* does not float, it can only be harvested at a good low tide.

CULINARY

Winged kelp has a lovely texture that, when reconstituted, has an almost noodle-like quality. Its mineral-rich ocean flavour has a subtle sweetness, and its savoury salty tones make it a lovely addition to soups, broths and chili. One of my favourite ways to use it is as a wrap for cooking fish. I soak a large piece of dried seaweed in cold water for several minutes, until reconstituted (if using dried winged kelp, it expands by about five times when reconstituted). Next, I lay half of the piece on a greased baking dish (or foil if barbecuing). I put a fillet of local fish on top, season, fold the remaining half of the seaweed over top of the fish, as if wrapping it in a blanket, then bake or barbecue. The seaweed keeps the moisture contained in the fish, and it becomes crispy, like chips, on top.

For many years, winged kelp was the only seaweed with which I could entice chefs. It truly is one of the best edible seaweeds, both in flavour and texture, and is well-loved as a culinary treasure. Closely related to Japanese wakame, the seaweed used to flavour the broth of miso soup, winged kelp is wonderful to use if making your own miso soup at home. Marinating it in lemon juice, which makes all the minerals even more bioavailable, and then slicing thinly and adding to salads is another yummy way to enjoy it. Cooking a piece of it with your rice helps to offset the calcium that is lost during the digestion of rice, and many say that adding it to beans will decrease the incidence of flatulence by preventing fermentation. For those who want the health benefits but have not yet acquired a taste for kelp, adding some ground flakes to a smoothie, or on top of a meal instead of salt, can be a great introduction.

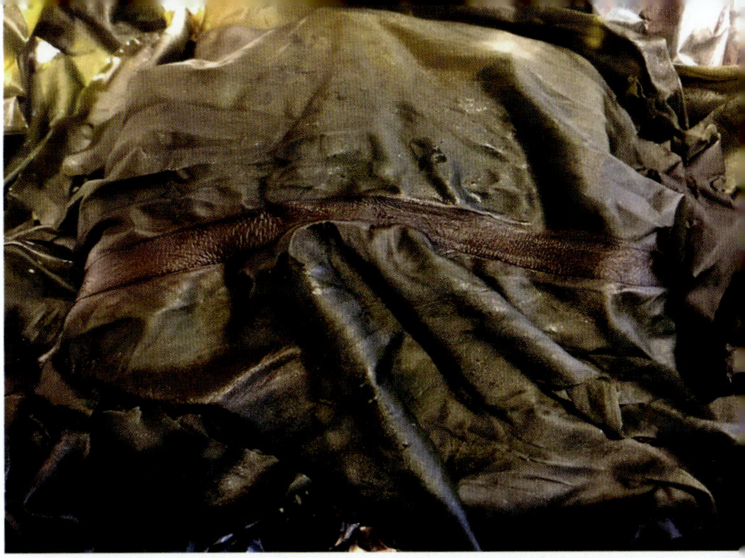

Local lingcod seasoned with grand fir and pepper, wrapped in winged kelp in preparation for grilling.

While previously seven local species of *Alaria* were described, phycological investigation has concluded, for now, that they all belong to a single species, *A. marginata*. This kelp is often called by its beautiful genus name, *Alaria*.

Ocean Style

The ocean has always enchanted me. Like an omnipotent god, there are true tales of her unbelievable, gentle mercy and her unimaginable power and fury.

I live beside the ocean and she never looks exactly the same twice. However, she does have *styles*. I find it fascinating how diverse her styles can be, and I am intrigued by each one.

Today she camouflages herself perfectly to match the grey sky. It is a style familiar to Pacific Northwesterners. With the persistent rain and low clouds, the difference in colour between dawn, day and dusk is difficult to decipher. On these days, the ocean pretends to be the sky, and you would never guess that she is a moving entity. You would never guess, if you had seen her dance with the wind, that she could be capable of such complete stillness.

Today she camouflages herself perfectly to match the grey sky. It is a style familiar to Pacific Northwesterners.

Sometimes you will see a wave approaching, moving like a snake along the perfectly still grey. You may, like I do, marvel at the electric splash of white as the wave breaks, completing its one-of-a-kind journey with its own death, never to exist again. If you are a rider of waves, you may feel both privileged to have seen this one and only ocean snowflake, and sad that you missed the one and only opportunity to slide down its liquid face before it disappeared forever onto the shore.

KATE WOODS PHOTO

Kombu KELP

Laminaria setchellii

Laminaria comes from the Latin word for "thin layer."

KINGDOM: Chromista
PHYLUM: Ochrophyta
CLASS: Phaeophyceae
ORDER: Laminariales
FAMILY: Laminariaceae

Several species are commonly referred to as "kombu," and they are all species in the genera *Laminaria* and *Saccharina*. Here in the Pacific Northwest, *Laminaria setchellii* is our local kombu. In Japan, kombu used to be reserved for royalty and the upper class, and it was given as a gift to bring about happiness. Most of Japan's kombu, which includes several varieties, comes from the northern Japanese island of Hokkaido. The highest quality kombu is luxuriously expensive and not available outside of Japan.

DESCRIPTION

Laminaria setchellii has a branched holdfast, a rigid, woody stipe and a deeply split blade with the splits occurring at approximately regular intervals. It is a classic kelp colour: golden brown mixed with olive green. *Laminaria setchellii* is a perennial seaweed that can live for up to twenty years and is typically less than one metre in length from the holdfast to the tip of the blade.

RANGE

Laminaria setchellii grows in the low intertidal and subtidal zones on wave-exposed rocky shores from Alaska to Mexico.

NUTRITIONAL PROPERTIES

Kombu is a rich source of vitamins A, B-complex, C, D, E and K and a rich source of minerals including calcium, magnesium, iodine, potassium, bromine, phosphorus and iron. It is an excellent source of fibre, protein and essential fatty acids. Kombu, specifically species in the *Saccharina* genus, contains the highest amount of iodine of any edible seaweeds.

AS A HEALING AID

Kombu has not only been a treasured food in Japan since ancient times, but it is also revered as medicine. The ailments that kombu has been

Laminaria setchellii with sea grass and red seaweeds draped over its woody stipe.

Dried strips of *Laminaria setchellii*.

traditionally used to prevent and treat would fill this page. These range from the prevention of colds and flus, to the treatment of cancer and mental illness, to the slowing of the aging process.

Kombu helps to safely detoxify the body of certain heavy metals and radioactive compounds, dioxins, PCBs and other toxins (see Toxin Cleansing, page 128). It is a very rich source of natural iodine, the highest of any seaweed by a significant amount, and can be very effective in the treatment of hypothyroidism. However, because it is so high in iodine, it should be used with caution by individuals with thyroid issues. In this case, working with a physician who can carefully monitor thyroid levels is recommended. Kombu is a rich source of biologically active compounds including fucoidan, fucoxanthin and phlorotannins. Studies suggest these compounds could be helpful in the prevention of certain types of cancers and chronic diseases (see Seaweed and Chronic Diseases, page 134); they also exhibit powerful antioxidant and anti-inflammatory activity. Kombu supports gastrointestinal health and cardiovascular health and promotes healthy brain function.

HARVESTING

I harvest very little *Laminaria setchellii*. It is a long-lived perennial seaweed, so I will harvest just a blade or two (technically divisions of the single blade) off any individual seaweed. I am careful to leave a significant part of the blade still attached. It is important to never cut the woody stipe, which would prevent the seaweed from regenerating. I primarily use *Laminaria setchellii* as a flavouring, so I consume much less than I do of a seaweed I snack on, like nori, dulse or sea lettuce.

CULINARY

Kombu contains the highest amount of glutamic acid of all seaweeds. Glutamic acid is a non-essential amino acid (meaning we can synthesize our own inside our bodies) required for protein synthesis and neurotransmission.

Smoked mussels with dashi broth, created by chef Andrew Dickinson. ANDREW DICKINSON PHOTOS

Japanese chemist Kikunae Ikeda isolated glutamic acid from kombu in 1908 and discovered that it was responsible for a taste sensation he experienced when eating *dashi*, a traditional Japanese broth made with kombu. This taste was different from the four basic tastes of salty, sweet, bitter and sour. He named this taste *umami*, which comes from the Japanese word for "delicious." Scientists have since identified specific receptors for glutamic acid, and umami is now considered one of the basic tastes. It is often referred to as the taste of savoury. Monosodium glutamate (MSG), the sodium salt of glutamic acid, is synthetically produced and sold as a flavour enhancer.

Kombu is exceptionally rich in iodine and contains an average of 3,000 micrograms per gram, though this amount is highly variable depending on the species, location and season. Cooking methods can also greatly influence the amount of iodine in kombu. In Japan, kombu is often boiled before eating to remove excess iodine. Kombu is best known for its use in flavouring dashi, a savoury broth that has been used as a foundational ingredient in Japanese cuisine for many generations. Kombu is also used extensively in Korean and Chinese cuisines, and in other coastal areas of East Asian countries.

Laminaria setchellii has a very thick, leathery texture. Its powerful umami taste enhances the flavour of other ingredients. It can be cooked with fish, beans and rice, and it is a wonderful flavouring in soups.

SPA

As *Laminaria setchellii* is a relatively small, long-lived perennial, I reserve the small amount I harvest for culinary uses. I choose larger, more abundant kelp for use in the bath.

The Phycological Society of America

The Phycological Society of America (PSA) hosts an annual international conference, which is a meeting where phycologists can present and share their current research and celebrate the study of algae. It is social, welcoming and adorned with seaweed art.

I attended my first PSA meeting in 2011 at the University of Washington in Seattle. There, I met Dr. Thomas Mumford, a phycologist studying the decline of bull kelp forests in Puget Sound, and Dr. Prannie Rhatigan, medical doctor, seaweed harvester, seaweed advocate and author of *Irish Seaweed Kitchen*. My friend and fellow seaweed enthusiast Christine Hopkins and I were so enamoured of her presentation, we invited her to give a seaweed workshop in our community the following year.

At the next PSA meeting I attended, in 2017, I was a presenter.

The meeting took place in beautiful Monterey, California. Though at that point I had been teaching and presenting for more than ten years, it would be my first time in front of an audience of seaweed experts. I was nervous, but also feeling excited about my presentation, which involved tasting and learning the nutritional and health properties of ten edible species of seaweeds I had harvested back home. Aside from Dr. Rhatigan's presentation in 2011, seaweed as food was a subject decidedly absent from the meetings. The thought of some of these phycologists, who had devoted their whole lives to studying seaweeds, actually eating some, was delightful to me.

One of the most incredible things I learned at the PSA conference in Monterey is that there are two genera of microalgae that dominate the world's oceans: *Prochlorococcus*, which dominates in nutrient-poor waters, such as in the middle of the ocean, and is represented by a single species (*P. marinus*) and *Synechococcus*, which dominates in nutrient-rich coastal waters. Anywhere you go in the world's ocean, if you take a sample of the water, you are likely to find one of these two genera dominating the population of microalgae (phytoplankton). That, by the way, was what phycologist Jeff Morris, from the University of Alabama, did when he wasn't in the lab: he collected samples of ocean water, by boat, from a huge area of the ocean.

Now I am aware, and proudly teach people, that half or more of the world's oxygen is produced by algae, and that most of this oxygen is produced by microalgae. Thus, after that talk, I wondered: If two genera of microalgae are the most abundant in the ocean and can be found almost anywhere in the ocean, does that mean that just two genera of microalgae are responsible for producing more of the world's oxygen than all terrestrial plants combined?

If two genera of microalgae are the most abundant in the ocean and can be found almost anywhere in the ocean, does that mean that just two genera of microalgae are responsible for producing more of the world's oxygen than all terrestrial plants combined?

Mind blown.

Adding even more potency to my revelation, I learned that *Prochlorococcus marinus*,

Louis Druehl speaking at the Phycological Society of America's 2018 meeting at the University of British Columbia in Vancouver. Seaweed artwork by Josie Iselin adorns the stage.

which NOAA reports is responsible for producing an estimated 20 per cent of the world's oxygen, is the smallest known photosynthetic organism on earth!

I relished soaking up the wealth of information offered by phycologists from far and wide and socializing with new friends in the evening, where the topic of conversation was always algae. I also discovered beautiful seaweeds, some known to me and some new to me, on the sunny beaches of Monterey. When your passion and field of study has such limited educational opportunities, it is easy to feel like a kid in a candy shop when you are surrounded by so many experts.

In one of these conversations, I learned that the genus name for nori, *Porphyra*, had been changed to *Pyropia*, and that sea lettuce was in fact two different genera: *Ulva* and *Ulvaria*. A phycologist who has been studying *Ulva* and *Ulvaria* for years confessed that with the naked eye, she can only correctly identify the genus she is looking at half the time when she is in the field. Both genera can grow in the same region and tidal zone.

During the break that day, I frantically opened up my presentation, which up until this point I had been calmly confident about, and began to try to "fix" some of the names. While I did this, my quirky sense of humour chimed in and I imagined saying something like this at my presentation: "Here are some edible seaweeds I have been harvesting near my home on Vancouver Island for many years now, but I really don't know what any of them are called."

Over some beers that night, I confessed my confusion about seaweed names to some young phycologists. One of them, a phycologist from the University of New Brunswick, where they are doing much of the DNA sequencing that is responsible for the name changes, replied: "Aren't we all. You are in some good company." He continued, "I hate to add to your stress, but the genus *Palmaria* is probably going to become two distinct genera in the next year or two. Genetically, Pacific *Palmaria* is totally unrelated to Atlantic *Palmaria*."

"Nooooo! Not *Palmaria*!" I protested.

I let out a sigh of defeat. Not even my precious dulse was safe. He went on to tell me what the new genus name might become for the current Pacific dulse (*Palmaria mollis*), which at least gave me the feeling of having a heads-up.

After that conversation, I felt reassured. The name game in the world of seaweed is dynamic. Better to just embrace it and be like the blades of bull kelp in a current: "go with the flow."

I went on to give my presentation the following afternoon. I was humbled and excited to share my edible seaweeds with such an illustrious group, and they warmly received the story of seaweed I had to share and the samples I had to taste.

The next PSA meeting was held at the University of British Columbia: my home turf of coastal BC. At the previous two PSA meetings I had attended, there had been hardly any mention of kelp forests. I thought to myself, "This meeting can't happen right here in BC, home of the world-famous and ecologically critical kelp forests, and not highlight kelp."

I was not disappointed. At PSA 2018, kelp was the star of the show, and the plight of kelp forests globally was highlighted as the most critical of concerns within the world of algae.

I have such admiration for this diverse group of phycologists. I am inspired to see many young phycologists so passionate about their work, often marrying phycology with current environmental concerns. The devotion to the pure, curious investigation of algae that is felt at a PSA meeting is infectious.

Louis Druehl and a student pulling farmed sugar kelp aboard the *Kelp Express* off the coast of Bamfield, BC.

I sometimes refer to Dr. Louis Druehl as the Kelp Guru. He is a pioneering kelp phycologist in the Pacific Northwest and studied under the renowned phycologist Robert Francis Scagel. He has published too many papers on kelp to list and is a professor of phycology. He is one of the founders of the Bamfield Marine Sciences Centre, and he was the first person to farm kelp in North America. In 1982, he and his wife, Rae Hopkins, founded Canadian Kelp Resources, which offers hand-harvested edible kelp. On a field trip during the 2018 conference of the PSA, I watched Louis, who was eighty-two years old at that time, pull a heavy line of his farmed sugar kelp aboard his humble skiff, the *Kelp Express*.

Egregia looks like patches of coarse hair sticking out of the water.

Feather Boa KELP

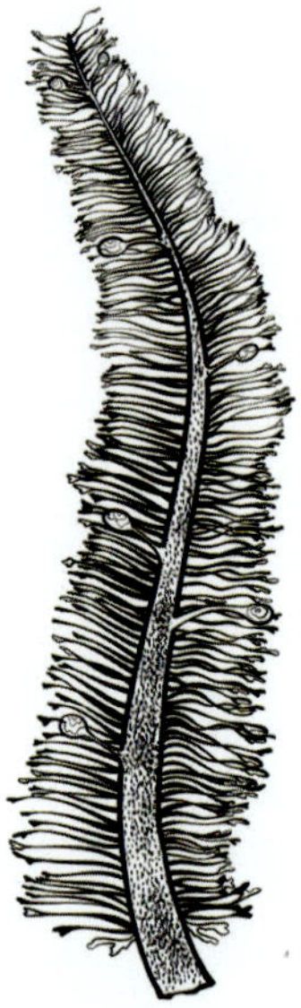

Egregia menziesii

Egregia comes from the Latin word *egregious*, which means "illustrious" or "excellent"; *menziesii* is named for the Scottish surgeon and botanist Archibald Menzies.

KINGDOM: Chromista
PHYLUM: Ochrophyta
CLASS: Phaeophyceae
ORDER: Laminariales
FAMILY: Lessoniaceae

DESCRIPTION

The common name for *Egregia menziesii* is "feather boa." If you get a chance to see this remarkable seaweed growing in its natural habitat, you will find it easy to imagine a mermaid wearing it around her neck at an underwater costume party.

Egregia is a large kelp that can reputedly grow up to fifteen metres in length, though I have never seen it longer than approximately five metres. It has many short blades, shaped like elongated teardrops, which grow laterally from a very thick and leathery midrib. Gas-filled bladders known as floats or pneumatocysts also grow

EMMA GEIGER PHOTO

laterally from the midrib, enabling the seaweed to float at the surface in the low intertidal and shallow subtidal zones, where patches of *Egregia* form a shallow water canopy. When viewed from the beach, patches of *Egregia* look like coarse hair sticking out of the water. *E. menziesii* is the only species in the genus *Egregia*.

RANGE

Egregia menziesii grows in the low intertidal and shallow subtidal zones from Alaska to Mexico.

NUTRITIONAL PROPERTIES

Egregia menziesii, a kelp, contains vitamins and minerals, but as it is not commonly eaten, there is little available information about its nutritional content.

AS A HEALING AID

As a kelp, *Egregia menziesii* aids in the safe cleansing of certain heavy metals and radioactive

Pneumatocysts and blades of *Egregia*.

The thick leathery midrib of *Egregia*.

compounds, dioxins, PCBs and other toxins due to its sodium alginate content (see Toxin Cleansing, page 128). The rich iodine content supports the thyroid.

HARVESTING

If you are planning to use fresh fronds, you can just trim some of them with a knife, leaving the leathery midrib intact. If you are planning to dry for later, it is easier to cut the midrib, leaving a good foot or two attached to the rock. This way, you can hang it to dry. Chefs and foodies often just want the floats to pickle—in this case, you can harvest the floats from several seaweeds, leaving the rest intact. *Egregia menziesii* becomes slimy when it gets older, so be sure you are harvesting seaweeds that are still vital. If you brush your hand against the frond and it leaves a brown slime on your skin, it is past its prime.

CULINARY

Chop the fresh blades and add to salads, soups and stir-fries. Enjoy dried fronds as a salty snack. Chefs love to play with pickling the floats. The most creative and aesthetic use of floats I have seen was as an "olive" in a "seatini."

SPA

Bathing with a large feather boa can really make you feel as if you are in the sea. It releases an abundance of nutrients into the warm, fresh water and permeates the air with the ocean's briny scent. It helps the body eliminate toxins and lactic acid, and it supports the thyroid. It is calming and soothing to the central nervous system and hydrating, toning and nourishing to the skin.

A Match Made in Kelp Heaven

The northern kelp crab, *Pugettia producta*, evolved legs perfectly designed to walk on kelp stipes. I see kelp crabs almost every time I harvest bull kelp. They hide in among the blades and they do indeed walk up and down the long stipes.

Kelp crabs are a relatively small crab. The largest one I have seen had a shell maybe four inches across. Their colour exactly matches the colour of the bull kelp. The older ones can have miniature ecosystems growing on their back with multiple seaweed species. I've seen a kelp crab with seaweed three times as long as itself growing off its shell. One time, I saw an entire bull kelp, complete with holdfast, stipe and a blade with a spore patch, growing off the shell of a particularly ancient-looking kelp crab.

When I took Montreal chefs Chuck Hughes and Danny Smiles out snorkelling and harvesting in the bull kelp forest as part of their six-episode documentary, *Chuck & Danny's Road Trip*, Chuck delighted me by taking a little nibble of a seaweed growing off the back of a kelp crab. The seaweed was *Costaria costata*, five-rib kelp. I imagined the menu write-up for such a unique ingredient: "five-rib kelp sustainably sourced from the shell of a living kelp crab."

When Chuck and Danny asked what we would see in the kelp forest, I told them some of the possibilities, such as jellyfish, lingcod, greenlings and sea lions, but said I could pretty much guarantee that we would see kelp crabs. I felt like one of those whale-watching outfits that promises sightings: "Dakini Tidal Wilds offers guaranteed kelp crab sightings, or your money back!"

Two kelp crabs with bull kelp blades showing the closeness of their colour.

Kelp crab on a bull kelp stipe in the kelp forest.

CHRIS ADAIR PHOTO

As all of my bull kelp apprentices are aware, I dedicate myself to many kelp crab rescue missions during the bull kelp season. As careful as we are to release them while we are putting the bull kelp blades into our bags out in the ocean, a few inevitably hitch a ride back to the seaweed shop. They usually make their break when the bag they are in is emptied in preparation for hanging: that's when they start their crabby sideways walk away from the safety of the bull kelp blades. Sometimes, though, they choose to stay hidden, and I have been known to give a surprised shriek when my hand reaches into the gelatinous pile of bull kelp and touches something with an unexpected texture.

Either way, they go into the bucket and I hop in my car and drive them to "the rescue beach," a beach just down the road from where we harvest, where I can drive right up close to the water. There, I walk the few steps down to the beach, out into the water, and with a farewell kiss and good wishes, I fling them toward the bull kelp forest off the western point of the beach. "Good luck, buddy," I usually say. "Enjoy your new home."

While my apprentice Geoff and I were hanging bull kelp after I returned from not one, not two, but *three* rescue missions, Geoff shared his theory about what a rescued kelp crab might say to his buddies about his experience upon being returned to the kelp forest, which I've paraphrased:

"Dude, I swear, I was abducted by aliens! One minute, I am chillin' in the blades; the next, I am still in the blades but the atmosphere is totally different! I make a run for it. I am cruising pretty good until one of the aliens, a total giant, picks me up and sticks me in a bucket. I figure I am done for at this point. Next thing I know, I am sailing through the air and all I can see is sky. I tell you, man, there are aliens out there."

Kelp crab with kelp and red seaweeds anchored to it.

Kelp crabs are so perfectly evolved to live a life interconnected with kelp that I often find myself lovingly thinking, while watching one cruising down a stipe in the kelp forest, "Now there's a match made in heaven."

Kelp crabs are so perfectly evolved to live a life interconnected with kelp that I often find myself lovingly thinking, while watching one cruising down a stipe in the kelp forest, "Now there's a match made in heaven."

JULY 2012

A Kelp to Call Home

In the kelp forest today, the place where my heart sings, I am struck very emotionally by the kelp crabs. In particular, I dwell upon their relationship with the bull kelp. I consider the depth of my own love for the bull kelp: the underwater forests they create and how they inspire in me a deep sense of purpose and belonging.

As I search the blades that I am harvesting for kelp crabs, relocating several via a combination of my hand and fin, It strikes me how deep and even seemingly *maternal* their connection is to the bull kelp. The bull kelp is their shelter, their world, like the earth is to us.

When I find a crab amid the blades that I am swooshing into my harvest bag as I negotiate the bouncy chaos caused by strong westerlies, I pluck them off the blade and they begin to descend, claws outstretched into the nothingness, to the sea floor far beneath. My heart breaking, I let go of the kelp I am busy loading into my bag and dive down, retrieve the crab mid-descent, deliver it onto the blades of another bull kelp, then return to the surface. I quickly gather the blades I have already cut, now dispersing with the wind and chop. Some beautiful pieces I retrieve just shy of the sea floor.

I have witnessed the relationship between kelp crabs and bull kelp for years, but for some reason today, seeing so many old ones lovingly nestled into the blades, and witnessing how exposed and vulnerable they appear when pulled away, I feel emotional in my contemplation. How nurtured and safe they seem, nestled in their mermaid-hair homes.

Me with a northern kelp crab, *Pugettia producta*.

AGATHE BERNARD PHOTO

it strikes me how deep and even seemingly maternal their connection is to the bull kelp. The bull kelp is their shelter, their world, like the earth is to us.

The branched holdfast, short stipe and puckered blade of *Costaria costata*. Notice that two of the midribs are convex (outward) folds and three are concave (inward) folds. If you flip the blade over, then three folds are convex and two are concave. EMMA GEIGER PHOTO

Five-Rib Kelp KELP

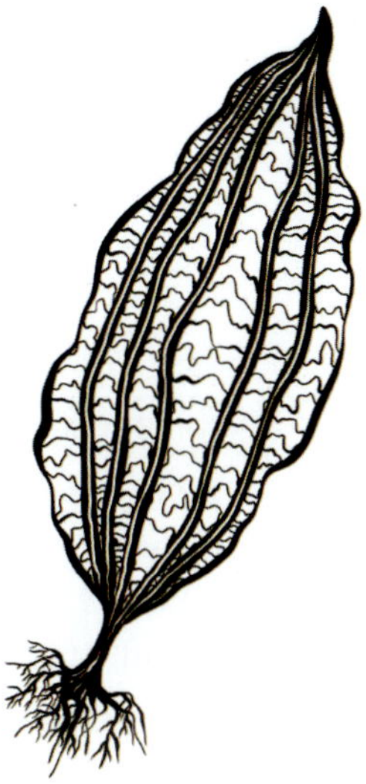

Costaria costata

Costaria comes from the Latin word for "rib."

KINGDOM: Chromista
PHYLUM: Ochrophyta
CLASS: Phaeophyceae
ORDER: Laminariales
FAMILY: Agaraceae

DESCRIPTION

Costaria costata is an impressive-looking kelp that can grow up to approximately three metres in length. It has a branched holdfast, a short cylindrical stipe and a single puckered blade with five midribs, which look like folds running the length of the blade. The puckered blade is reminiscent of the back of an alligator, and when I teach, I sometimes tell kids that another name for five-rib kelp could be "alligator kelp." *Costaria costata* displays phenotypic plasticity, which means it can alter its morphology depending on the environmental conditions of its location. For example, in wave-exposed areas, *Costaria costata* grows a long, slender blade, and in more sheltered areas, it grows a shorter, broader blade. The long and slender blade protects the seaweed against being ripped up due to wave action, and the broader form maximizes photosynthesis.

RANGE

Costaria costata grows in the low intertidal and subtidal zones from Alaska to central California.

NUTRITIONAL PROPERTIES

Costaria costata is a rich source of vitamins A, B, C and E and minerals including potassium, magnesium, calcium, iron and iodine. It is a source of protein and an excellent source of essential fatty acids and fibre.

AS A HEALING AID

Like all kelp, *Costaria costata* helps to safely detoxify the body of certain heavy metals, radiation, dioxins and PCBs. It contains biologically active compounds including fucoidan, fucoxanthin and

COURTNEY STANDING PHOTO

Costaria costata restores moisture and vitality to the hair and skin. Its abundance of nutrients, combined with the pure ocean aroma it diffuses into the air, induces calmness, helping to combat stress.

phlorotannins, which exhibit cardioprotective, neuroprotective, anti-inflammatory, antioxidant, antibacterial, antiviral, antifungal and anticancer activity. It is a rich source of iodine, nourishing the thyroid and supporting metabolic health. (See Seaweed as Protector, page 128, and Seaweed and Chronic Diseases, page 134.)

HARVESTING

I harvest *Costaria costata* in the spring and early summer. If I am harvesting for food, I will cut the blade of a young *Costaria* several inches above the stipe. If I am harvesting for use in the bath, I will salvage larger blades that are still fresh and vital. I don't harvest very much *Costaria costata* for eating, as I prefer the flavour of other kelp, but I love to have large pieces to soak with in the bath.

CULINARY

Costaria costata can be used to create nutritious, mineral-rich broths with a salty umami flavour. On its own I find it more bitter than other kelp such as bull kelp and winged kelp.

SPA

Bathing with a large piece of *Costaria costata* is a wonderful experience. It permeates the air with the scent of the sea, and watching it float can make you think there is an alligator in the tub. When exposed to fresh, hot water, the colour instantly changes from golden brown to bright olive green, and you can feel the smooth, healing alginates seeping out. *Costaria costata* encourages the elimination of toxins and lactic acid through the skin, delivers vital nutrients to the skin and hair and promotes a feeling of well-being and relaxation.

SEPTEMBER 2011

Under the Ocean Canopy

It is a cold mid-September day, and I am doing a solo mission, like old times. It is my youngest daughter's second full day of kindergarten and taking the bus. It is the first time that I am at home all day while both of my daughters are in school. It feels very weird. It is also the first day of the rainy season. The period of persistent low-cloud, timeless light and steady rain, all too familiar to inhabitants of the Pacific Northwest, is starting.

After taking down all the dry bull kelp inside the shop from the harvest two days ago, I suit up and head down to the beach, my favourite place in the world. As I zip up, don my mask, fins and snorkel and head into the grey unknown, the words of Andrew Princz, a journalist who had interviewed a seaweed harvester, appear in my thoughts: "If writing is a lonely task, I thought, her daily routine of plucking seaweed from this ocean playground must be equally forlorn."

The water is cold today. The water is always cold on the west coast of the island, but today it is deeply icy. It is so full of floating nutrients that it appears white, like snow, on a backdrop of green the exact colour of glaciers. As I meld with the water, my senses expand, and my mind sinks into the slow, all-consuming sound of my breath under water. The cold pierces me as I breathe into its energy and allow it to enter me.

Seeing large patches of dulse, stringy *Desmarestia* and the occasional walking kelp, I also see underwater landmarks that I have observed for nine years. I feel blessed, like I always do when I am swimming out toward the open sea. For nine seasons, every time, *every single time* I swim out to these kelp forests, I feel the presence of something deeply spiritual. To borrow the words of Bob Marley, I feel a "natural mystic." Jess, who has been apprenticing with me all season, hears me proclaim each time: "This was honestly the most amazing and beautiful I have ever seen it!" This could sound exaggerated, untrue, but she understands just what I mean because she experiences it too. It is alluring, enchanting and different every time.

As I meld with the water, my senses expand, and my mind sinks into the slow, all-consuming sound of my breath under water. The cold pierces me as I breathe into its energy and allow it to enter me.

Today, for the first time ever, I see a bell-headed tailed jelly (*Praya* spp.). It is definitely one of the stranger things I have seen in my life. I inspect what appears to be a collection of peach-coloured, oval-shaped beads, spread out in a long line, held together by a nearly invisible, clear, gelatinous casing about one foot in length. When I read up on the species later that afternoon, I learn that they can grow up to fifty metres long and are one of the longest animals in the world!

The bull kelps look like sentinels and soon become so dense that one can only weave

Kelp crab on bull kelp stipes in a kelp forest.

CHRIS ADAIR PHOTO

between them under their canopy, which is where I love to be.

Under the canopy is a whole other world. There are ancient-looking rockfish, kelp greenlings and sly-faced lingcods hovering against the seaweed-covered bottom. Large schools of Pacific sand lance and surf smelt weave as a single entity around the stipes of the bull kelp beneath the thick canopy, which blocks out most of the sunlight from above. I have even seen the odd full-grown salmon lurking in the shadows.

Kelp crabs abound, travelling along the bull kelp stipe highways. Diving down underneath the canopy, feeling the pressure of held air against my chest, eyes wide in wonder at the teeming variety of life that exists down here, the stresses of my life on shore quickly dissipate. The space they filled inside me gives way to wonder and enchantment.

Bull kelp connects surface and depth, encompassing both. It also has a superpower: resilience. In an adaptation to help withstand the force of swell and surge, the stipe can stretch by as much as 30 to 40 per cent of its length.

Perhaps there are lessons to be learned from bull kelp: be flexible enough to withstand life's stresses, grow toward the sun at the surface, but always keep your holdfast anchored deep.

CHRIS ADAIR PHOTO

Triple-Rib Kelp KELP

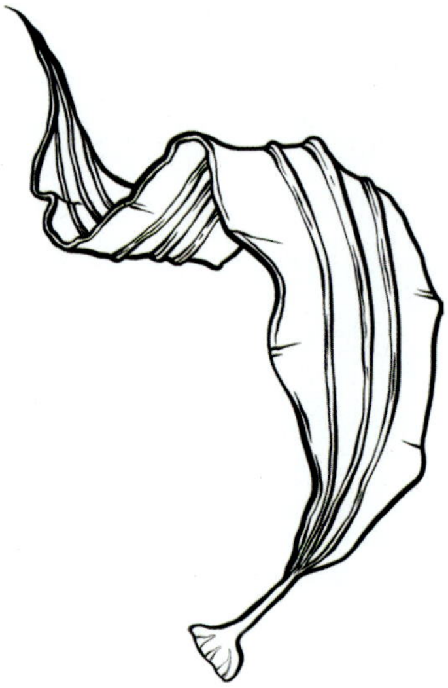

Cymathaere triplicata

Triplicata is Latin for "three folds."

KINGDOM: Chromista
PHYLUM: Ochrophyta
CLASS: Phaeophyceae
ORDER: Laminariales
FAMILY: Laminariaceae

DESCRIPTION

Cymathaere triplicata is an annual seaweed that has a single blade, a short cylindrical stipe and a discoid holdfast. The common name, triple-rib kelp, is slightly misleading. It appears to have three ribs running the length of the blade in the centre, but they are actually folds. This kelp can grow up to approximately three metres in length.

RANGE

Cymathaere triplicata grows in the low intertidal and subtidal zones from Alaska to Washington.

NUTRITIONAL PROPERTIES

Triple-rib kelp is an abundant source of vitamins and minerals. Because it is not a common edible, specific nutritional information is not available.

Cymathaere triplicata with sun rays.

AS A HEALING AID

Like all kelp, *Cymathaere triplicata* contains sodium alginate, which helps to detoxify the body, and biologically active compounds such as fucoidan and fucoxanthin, which studies suggest help to prevent certain types of cancers and chronic diseases.

HARVESTING

I harvest *Cymathaere triplicata* in the late spring and early summer. As an annual, it is not very large until later in the spring. I harvest at a very

Cymathaere triplicata, identifiable by its distinctive three folds, attached to the same rock as *Alaria marginata*.

EMMA GEIGER PHOTO

low tide and leave several feet of this large kelp attached. I also harvest by salvaging, often finding fresh long blades washed up on my beach while walking my dog.

CULINARY

Fresh *Cymathaere triplicata* has a crisp texture and tastes like a cross between a cucumber and an olive. This umami-rich kelp has a sweet and tangy flavour and imparts a tingling sensation to the outer sides of the back of the tongue.

SPA

Triple-rib kelp is therapeutic when bathed with, helping to bring about calm, eliminate toxins and tone and nourish the skin. Just a small piece of dried triple-rib kelp will fluff up into a large, gelatinous frond, expanding by about five times when submerged. It will also add the aroma of the sea to your bathing experience. It releases nourishing vitamins, minerals and alginates into warm, fresh water and the integrity of its texture allows it to be used multiple times if it is dried between baths.

SEPTEMBER 2012

Fish School

It is a brisk morning: sun shining, ocean glassy, no current and a nice low tide with tiny swell rolling onto the shore: my favourite conditions for experiencing the beauty of the kelp forest.

Swimming out toward the kelp forest, I marvel at the corridors of sunlight dancing on the sandy bottom and illuminating some of the reds like a spotlight. A large school of sand lance, swimming with undulations like eels, heads toward the shelter of the kelp forest. When I arrive at the forest's edge, it is beautiful beyond imagination.

When the conditions are like this, I am truly swimming in an emerald sea. Diving under, the water glows a whitish green, the colour of arctic ice. The bull kelp fronds, the mermaid's hair, hang down toward the ocean floor. Large schools of small fish feed on the abundant plankton suspended in the water. As I drift, they don't pay any attention to me: I am just part of the plankton, moving with the ocean currents. However, as soon as I move a fin, just the tiniest push, creating my own current, the fish instantly organize themselves and swoosh this way and that, their movements appearing to be those of a single creature. Though I don't like to disturb them, I sometimes can't resist diving into the centre of their school for the ultimate psychedelic experience of being immersed in a three-dimensional, dynamic tunnel of silvery fish. Most of the time, though, I just stare in fascination at their dance of movement: hundreds of individual creatures flowing rhythmically as one. I reluctantly pull myself away from their fascinating display and begin to harvest blades of bull kelp.

Immersed in a school of fish in the darkness beneath the canopy of a kelp forest. CHRIS ADAIR PHOTO

Midway through my harvest, I peer past a cluster of entwined bull kelp stipes and see the largest school of fish I have seen all season. This massive school stretches from the surface to the very bottom, fifteen to twenty feet deep, and as far as I can see in three directions. The stipes of the bull kelp stand like pillars, the school of fish

Large schools of small fish feed on the abundant plankton suspended in the water. As I drift, they don't pay any attention to me: I am just part of the plankton, moving with the ocean currents.

Taking a break from harvesting to be amazed by a school of silvery sand lance. CHRIS ADAIR PHOTO

parting to make way around each one. As I hang weightless in the water, they ease their tight formation. The single entity returns to thousands of individuals, each eating plankton in a motion that makes them appear to be kissing an invisible mate. The fish, Pacific surf smelt, are perfectly uniform: silver with hues of pale pink, yellow and turquoise that can only be seen if you are within inches of them.

I finish filling up my harvest bag and hover at the surface for a minute. A small school of five or six tube-nosed poachers comes within inches of my mask. They are small, curious-looking fish, reminiscent of sea horses.

While snorkelling back to shore, I see a small, flat fish. It is a beautiful sanddab, partially covered in sand and perfectly camouflaged. I wouldn't have noticed it if I hadn't dived down and been just inches away. Its body sparkles just like sand does in the sun.

I reach the shore, drop my load of seaweed into my wheelbarrow and push it up the hill to my 4Runner. I drive the couple hundred metres back to my place and prepare to hang the blades of the bull kelp in my drying shop.

There is an inimitable healing energy to be found in the kelp forest. Today, it was fish medicine. Star-bright flashes of sparkling silver. The unexpected union of separateness and wholeness. Immersion in a school of fish thousands strong is an experience that is sacred in this world. Fish school magic.

Sea Cabbage KELP

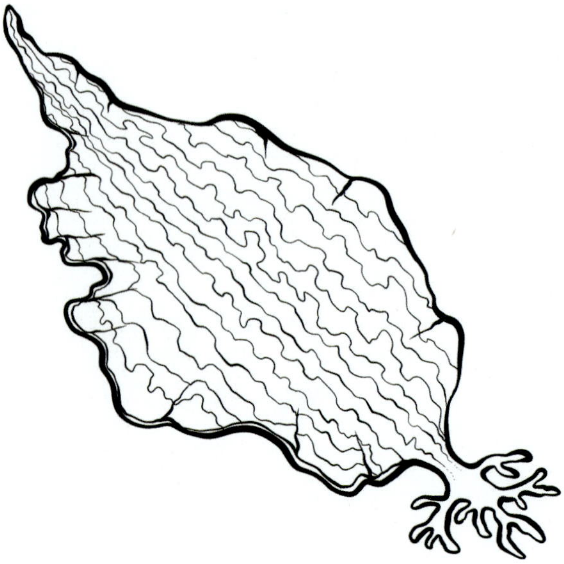

Saccharina sessilis (formerly _Hedophyllum sessile_)

Saccharina sessilis comes from the Latin words for "sugary" and "immobile."

KINGDOM: Chromista
PHYLUM: Ochrophyta
CLASS: Phaeophyceae
ORDER: Laminariales
FAMILY: Laminariaceae

DESCRIPTION

Saccharina sessilis is the simplest of the kelp. It lacks a stipe and has blades that are bullate (blistered or puckered) and grow directly from a holdfast composed of many haptera. Sea cabbage is small for a kelp and its blades are less than one metre long. In very exposed areas, the blades are short, smooth and slender and have many divisions that look like rips. In sheltered areas, the blades are longer, wider, less divided and more bullate.

RANGE

Sea cabbage grows in the mid intertidal zone from Alaska to central California.

NUTRITIONAL PROPERTIES

As a seaweed and kelp, sea cabbage contains vitamins, minerals and other nutrients. However, as it is not commonly eaten, specific nutritional information is not available.

AS A HEALING AID

Like all kelp, sea cabbage contains sodium alginate, which helps to safely cleanse certain heavy metals, radiation and other toxins from the body (see Toxin Cleansing, page 128). It is a source of fucoidan and fucoxanthin, which research suggests are helpful in the prevention of certain types of cancers and chronic diseases such as CVD and type 2

Sea cabbage anchored to a rock at low tide.

diabetes, and which exhibit anti-inflammatory and antioxidant activity (see Seaweed and Chronic Diseases, page 134). Sea cabbage also hydrates and nourishes the skin.

HARVESTING

I harvest sea cabbage in the spring and early summer by pruning the blades. The distal end of the blade, which is farthest from the new growth, is often in a state of deterioration, particularly in wave-exposed areas, so I trim this part off. I harvest very little, as I usually prefer other species in my kitchen cupboards.

CULINARY

Sea cabbage has a texture like land cabbage when eaten raw. However, the alginates are quite prominent when chewed, and the gelatinous texture can be unpalatable. It also imparts a metallic taste on the tip of the tongue. Sea cabbage dries nicely and can be chopped up and added to soups, where the rich alginates act as a thickener.

SPA

Sea cabbage releases abundant alginates in the bath, which encourage the elimination of toxins and lactic acid and help to heal, nourish and tone the skin. For maximum benefit, rub the seaweed directly onto the skin.

In semi-sheltered areas, blades of *Saccharina sessilis* are bullate (blistered), and in wave-exposed areas they lack much of this morphological feature.

Elegance at Work

The water is perfectly still, a mirror to the sky. I feel that amazing feeling I sometimes do when I am harvesting alone: that I am a natural part of the intertidal environment. Wading around in the still waters of the Juan de Fuca Strait, I instinctively dip my hand under water to cradle the long, smooth fronds of *Alaria*, with their golden midribs. Deep in my harvest meditation, I am gifted an intimate experience with a great blue heron.

He flies over my head and lands very near to me, close to the shore. He begins to fish in the shallows with admirable concentration. He looks over at me a few times, and each time I find myself looking away shyly—embarrassed that he has caught me staring at him in wonder—and busy myself with my harvesting. I feel deeply humbled by his elegance.

Toward the end of my harvest session, he lifts off into flight. He is the most skilful flyer I have ever seen. The shapes he makes with his neck are fluid, and I can see each movement as if he is flying in slow motion. I watch as he lands farther down the beach, intently scanning the new location for fish. Though he is a very large bird, he lands with the deliberateness and grace of a ballerina coming out of a pirouette.

He is the most skilful flyer I have ever seen. The shapes he makes with his neck are fluid, and I can see each movement as if he is flying in slow motion.

I have seen lots of great blue herons, but this experience was different. It was the first time it felt like we were both natural creatures of a shared habitat, existing side by side in the intertidal zone. He was aware of my presence and fished alongside me: he for fish and I for seaweed.

KATE WOODS PHOTO

Rockweed

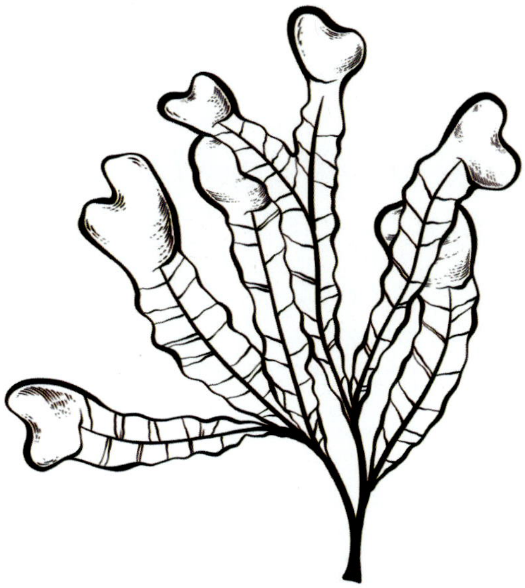

Fucus distichus* (formerly *Fucus gardneri*) and *Fucus spiralis

Fucus is derived from the Greek word for "seaweed."

KINGDOM: Chromista
PHYLUM: Ochrophyta
CLASS: Phaeophyceae
ORDER: Fucales
FAMILY: Fucaceae

DESCRIPTION

Fucus distichus and *Fucus spiralis* are olive-green to golden-brown seaweeds with small discoid holdfasts and flat, dichotomously (Y-shaped) branched blades with midribs. They can grow to a maximum of approximately forty centimetres. In the field they are easily recognized by the warty-appearing terminal bladders often referred to as "poppy things." These poppy things are called "receptacles" and contain the seaweed's egg and sperm. If popped, they release a liquid similar in texture and consistency to aloe vera gel. This gelatinous liquid is rich in alginates and fucoidan and is UV resistant in order to protect the gametes from light damage.

It has recently been discovered that the seaweed *Fucus gardneri* is the same species as *Fucus distichus*, and consequently the species name *F. gardneri* has been dropped. They were long believed to be different species due to the different regions they grow in: *F. distichus* was associated with growing in the North Atlantic and *F. gardneri* in the North Pacific. Some phycologists consider them subspecies, due to the different regions where they grow. In the Pacific Northwest as well as some other regions, *Fucus* species are known as rockweed.

Seaweeds belonging to the order Fucales are widely distributed, growing on most temperate coasts across the world. They contain an abundance of alginates, fucoidan, fucoxanthin and other biologically active compounds that show great potential in promoting human health. They possess many similarities to kelp (seaweeds belonging to the order Laminariales), and together are the most widely studied group of seaweeds for their medicinal properties. Phycologists believe that seaweeds belonging to the order Fucales are the ancestral predecessors to kelp.

RANGE

Fucus species inhabit most cold-water coasts around the globe. In the northeastern Pacific Basin, *Fucus distichus* occupies the mid and high intertidal zones, and *Fucus spiralis* grows in the high intertidal zone on sheltered and semi-exposed rocky shores from Alaska to central California.

NUTRITIONAL PROPERTIES

Fucus species are an excellent source of vitamins A, C, E and K and are abundant in essential minerals including iodine, calcium, magnesium, bromine, phosphorus, zinc, iron and potassium. *Fucus* species are excellent sources of fibre and essential fatty acids and are abundant in biologically

Fucus distichus exposed at low tide.

active compounds including alginates, fucoidan, fucoxanthin and phlorotannins.

AS A HEALING AID

I sometimes call *Fucus* species the "dandelions of the sea." Like dandelions, they are often very prolific, with members of the Fucaceae family growing on most temperate coastlines; also like the common dandelion, *Fucus* species have phenomenal health benefits. Most "kelp" tablets and powders I have seen sold as supplements in health food stores are not kelp at all, but some species of *Fucus*, most often *Fucus vesiculosus*. From a healing perspective, this is okay because *Fucus* species share a lot of properties with kelp. Like kelp, they are rich in iodine, sodium alginate, fucoidan, fucoxanthin and phlorotannins.

Fucus species encourage the safe elimination of certain heavy metals, radiation, dioxins, PCBs and other toxins and have anti-inflammatory and antioxidant activity (see Toxin Cleansing, page 128, and Radiation Protection, page 128). *Fucus* species are abundant in fucoidan and fucoxanthin, which research suggests can aid in the prevention of certain cancers, CVD, type 2 diabetes and other chronic diseases, in addition to enhancing immune function (see Seaweed and Chronic Diseases, page 134). *Fucus* species are helpful for thyroid disorders associated with hypothyroidism, and they can help to alleviate the side effects of radiation therapy and chemotherapy. The gel-like substance in the receptacles can be used as a natural sunscreen and to soothe, cool and promote healing of sunburnt skin.

Fucus vesiculosus, which grows in the North Atlantic and is known by the common name "bladderwrack," has been used for generations in Western herbalism. It is a traditional remedy for arthritis and joint pain, anxiety, heartburn and constipation, urinary tract infections, respiratory ailments, tuberculosis and thyroid disorders including goitre. It has also been used for immune support and metabolic support, and to ward off colds and flus. It contains significant amounts of

Harvesting *Fucus* at low tide. EMMA GEIGER PHOTO

Fucus distichus with sea lettuce and red seaweeds. EMMA GEIGER PHOTO

bromine, which studies show has a toxic effect on tuberculosis-causing bacteria. Topically, it has been traditionally used to relieve burns, insect bites and rashes.

NOTE: *Fucus* species should only be used *following* radiation treatment, not during, due to their effectiveness at removing radiation from the body. The nutritional benefits of seaweed are provided by consuming a small amount regularly, about five grams per day. If you plan to use *Fucus* species or other seaweeds therapeutically, be sure to consult with a medical professional first.

HARVESTING

I am extra conscious when harvesting *Fucus distichus*, taking only a few strands from each patch and harvesting only in areas where there is an abundance. While this level of consciousness is required for sustainably harvesting any seaweed, it is especially true when you are selecting the reproductive part. Sodium alginate and fucoidan are concentrated in the receptacles (the bladders at the terminal ends of some of the branches, which contain the seaweed's egg and sperm), so they are desired, though the whole seaweed is useful. Like always, leave a good amount of seaweed attached to the rock. I harvest *Fucus distichus* in spring and summer.

Submerged *Fucus* species with floating tips.

Fucus species are my favourite seaweeds to use in foot baths. They turn an electric green when exposed to hot fresh water and release an abundance of healing alginates and other nutrients.

CULINARY

I seldom eat rockweed, preferring to use it in the bath, but when I do, this is my favourite way to prepare it: Heat up a skillet on medium-high heat, and when the skillet is hot, add a little bit of olive oil, minced ginger, balsamic vinegar and tamari. Add fresh rockweed tips and cook for about forty-five seconds, until the tips turn an electric green, then remove immediately and serve. The goal is to crisp the outside while leaving the inside, which contains the alginates and the fucoidan, raw. If you cook it too long, you will end up with a pile of very gooey seaweed that few people will find palatable.

Rockweed is quite mild flavoured compared with other seaweeds. I sometimes offer folks with a brave palate a fresh tip of rockweed on my seaweed tours. Though the rockweed tips can look a little intimidating, participants are usually surprised at how mild they are, and often try to convince their less palatably brave spouse to try one.

My friend and colleague Christine Hopkins takes advantage of the gelatinous nature of rockweed tips to make a glaze for her locally famous focaccia bread, grinding up the tips and mixing them with olive oil, minced garlic and parmesan cheese.

SPA

Fucus species have tremendous therapeutic value, and their use in thalassotherapy (bathing in seaweeds) in Ireland and other European regions stretches back centuries. Foot baths, hand soaks, full baths and steam inhalation are all very therapeutic. Thalassotherapy can help with respiratory ailments, sore and inflamed joints, circulation, regulating metabolism, soothing and calming the central nervous system and balancing the endocrine system, and it is deeply nourishing to the skin. *Fucus* species gently remove toxins and are rejuvenating and moisturizing.

Other Common Brown Seaweeds of the Pacific Northwest

DESMARESTIA HERBACEA >

Desmarestia herbacea is one of the bladed forms of the *Desmarestia* genus. The common name of bladed *Desmarestia* species is "flattened acid kelp." This common name is partly accurate: *Desmarestia* species are not kelp, but species in this genus contain sulphuric acid. They should definitely not be consumed, and they will spoil other seaweeds if collected in the same bucket or bag. There are bladed forms of *Desmarestia* species, such as in the photo, and cylindrical forms that have a stringy appearance.

EMMA GEIGER PHOTO

< PLEUROPHYCUS GARDNERI KELP

Pleurophycus gardneri, a kelp, is common on wave-exposed rocky beaches in the Pacific Northwest. It looks similar to *Alaria marginata*, winged kelp, and has the common name "broad-rib kelp."

POSTELSIA PALMAEFORMIS KELP >

Postelsia palmaeformis looks like a seaweed version of a palm tree, earning it the common name "sea palm." I find this kelp one of the most intriguing in appearance. Even though it is common on the west coast of Vancouver Island, it only grows on the most exposed shorelines where, unlike most other seaweeds, it can endure being pounded by huge waves. For many years after my seaweed journey began, the only place I saw sea palm was in books, because my local shores are only semi-exposed, and I longed to see it in the flesh. On a trip to Oregon, I was absolutely thrilled when I found this one washed up on the fully exposed beach I was walking along. I was later fortunate enough to see a rock offshore that was exposed by a low tide and thickly covered with sea palm. I was astounded at how this beautiful, tree-like kelp could remain attached and healthy as head-high waves forcefully crashed into it.

< *SARGASSUM MUTICUM*

Sargassum muticum is an introduced species from the northwestern Pacific Basin. It can grow into prolific beds and frequently outcompetes native species, especially in sheltered regions. For this reason, it is considered an invasive species.

SCYTOSIPHON SPP >
The cylindrical brown seaweeds in this genus consist of hollow tubes, some of which have defined sections. Perhaps "miniature sausage seaweed" would be a good common name for this interesting-looking alga?

< *SORANTHERA ULVOIDEA*
This small, globular brown seaweed is covered with bumps.

SEAWEED & HEALTH

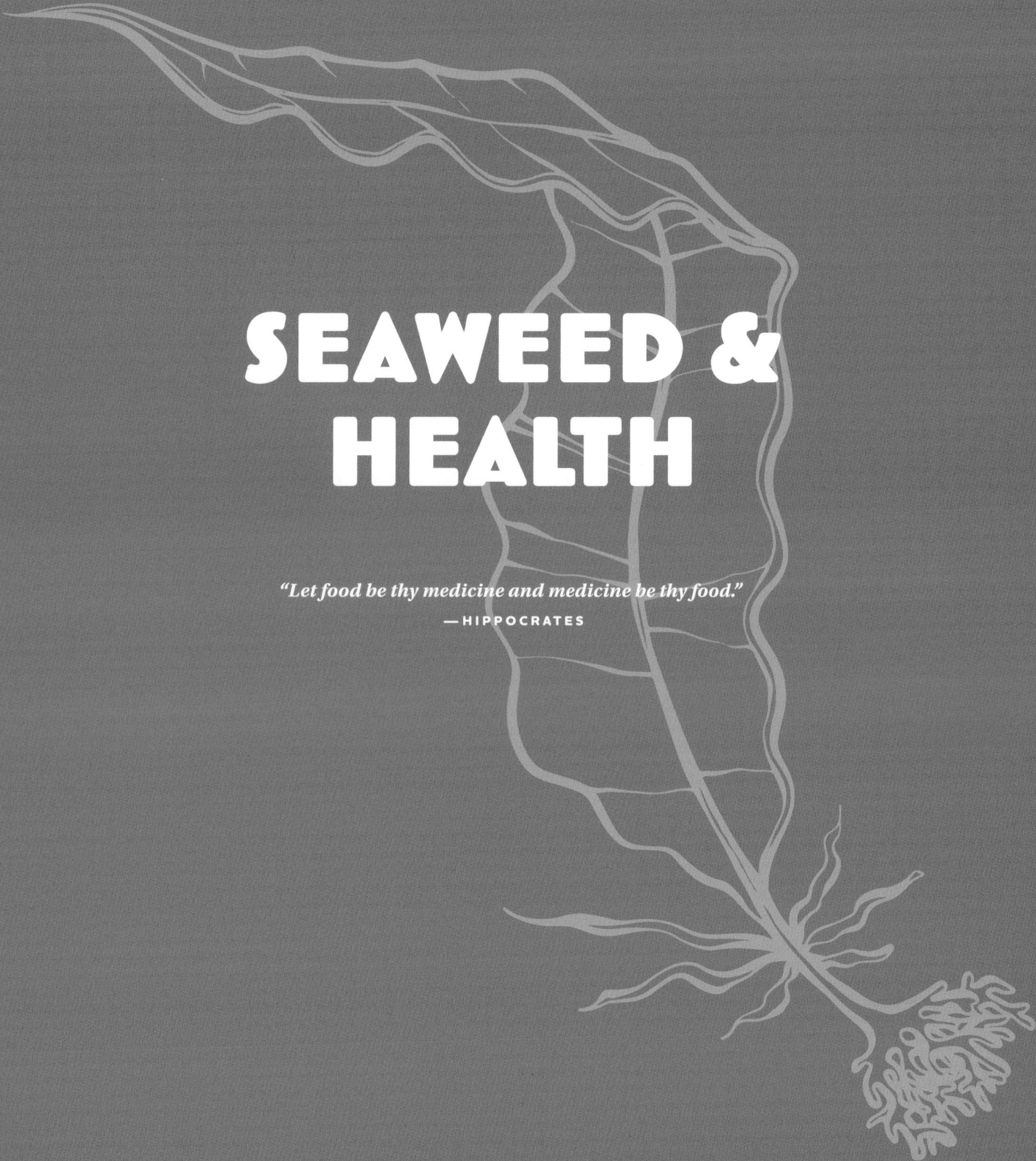

"Let food be thy medicine and medicine be thy food."

—HIPPOCRATES

Ho'oponopono

Limu is the Hawaiian word for "seaweed" and *kala* is the Hawaiian word for "forgiveness." The Hawaiian seaweed limu kala (*Sargassum echinocarpum*) is a small seaweed belonging to the brown seaweed group and has spikes along its margins. It is used in the Hawaiian forgiveness and reconciliation ceremony called Ho'oponopono. When people are in conflict, they gather with their families, including those not involved in the conflict. A circle is formed, and all members remain in the ceremonial circle until harmony is restored among the people in conflict, whether that takes a few hours or a few days. Limu kala is used as a purification tool and is symbolic of release and reconciliation.

Limu kala. KATHLEEN HO PHOTO

The Ho'oponopono prayer has been adapted by people throughout the world and is currently used as a meditation and spiritual practice. It can be recited in any order.

THE HO'OPONOPONO PRAYER

I'm sorry
Please forgive me
Thank you
I love you

Ancient Medicine from the Sea

Over thousands of years, seaweeds have gained a rich history as a medicinal food. They have been used to ward off sickness, infection and chronic diseases, and to promote vitality, beauty and longevity. When I began harvesting seaweeds, I had no idea of the absolute wealth of healing potential they possess. Thousands of studies have examined the properties of seaweeds and seaweed extracts and how they affect human health, with many more happening currently.

Seaweeds are the most concentrated food source of minerals on the planet. They store these minerals in chelated and colloidal forms, rendering them easy to absorb and bioavailable. Minerals are major components of our organs, tissues, glands and foundational cells, and an optimal ratio of minerals in our blood is critical to our survival. Seaweeds are rich in the essential major minerals: sodium, potassium, chloride, calcium, magnesium, phosphorus and sulphur. They are also an excellent source of essential trace minerals including iron, zinc, iodine, chromium, selenium, manganese, copper, fluoride and molybdenum. Absorbing minerals directly from the ocean, seaweeds have evolved over millennia to have an optimal ratio of minerals. I think of them as nature's perfect mineral supplement.

Of particular importance is seaweeds' abundance of natural iodine, an essential mineral that an estimated two billion people worldwide are deficient in. Iodine is a major component of the human thyroid gland, which regulates metabolism and is associated with protein synthesis, children's growth and development, lactation and other vital functions. Iodine deficiency is reported to be the leading cause of preventable intellectual disability.

Seaweeds are among the world's most nutritious foods, packed with vitamins, minerals, essential fatty acids, fibre, protein and unique biologically active compounds.
JESSE ORRISS PHOTO

Seaweeds are also among the highest food sources of vitamins. Most edible kinds contain vitamins A, B-complex, C, D, E and K, as well as many complementary elements required for the optimal absorption of these vitamins.

Seaweeds are high in both soluble and insoluble fibre, now known to be a critical factor in gastrointestinal health and the prevention of chronic diseases such as CVD and type 2 diabetes. Seaweeds are a good source of protein, too, some containing as much as 45 per cent (dry weight), and they are also a source of essential fatty acids, including significant amounts of omega-3s. They also contain a wealth of unique biologically active compounds not found anywhere else in nature. Some of these compounds offer astounding benefits to human health.

Seaweeds themselves are known to grow in some of the harshest environments on earth and not only survive but thrive. In particular, seaweeds growing in the intertidal zone are exposed to desiccation, extreme variations in temperature and salinity, wave action, a plethora of both airborne and marine bacteria, viruses and fungi and, in colder regions, freezing and thawing. Their adaptions to these extreme living conditions have resulted in a medley of novel therapeutic substances exclusive to seaweeds.

Since seaweeds are such a nutrient-dense food, very little is required to receive the myriad health benefits. Eating a small amount, just a few grams per day consistently, is the best way to support and enhance health. Bathing in seaweeds is another way to instill benefits. The skin is an efficient way to absorb many of seaweed's nutrients, and bathing in seaweed can offer different therapeutic benefits than consuming it does. Seaweed baths have been used for millennia to promote an overall feeling of well-being, in addition to many other therapeutic applications.

I feel a responsibility to share with others what I have learned about the protective and supportive qualities of seaweeds. This section is an invitation to contemplate these findings, and to investigate further.

MARCH 2019

Seaweed Therapy

It is early in the morning. My daughters are still asleep, and the house is quiet. I look out the window to see daylight gently arriving. In another month, a new seaweed harvest season will begin. I contemplate the work and life I have chosen.

I have been a seaweed harvester for seventeen years now, coming up on my eighteenth season. It has become a meditation for me, a practice. It is the thing I know how to do. During the peaks and valleys of nearly two decades of life, it is the thing that has remained constant. When the tide is low, no matter what else may be going on in my life, I am out harvesting seaweed.

The kelp forest and seaweed gardens of the intertidal zone have become the place I navigate my life: a place tranquil enough for me to make some sense of the big, scary, beautiful world. When I think of all I have brought out with me to the place where seaweed thrives—the devastating losses, the joy-filled triumphs and everything in between—it makes me feel so much.

Gratitude. I am so passionate about what I do. I am in the ocean every day, wildcrafting, a part of a dynamic food community. My work is self-determined, health-giving and reliant on trusting relationships. It has a soft environmental footprint, serves the community and inspires me to continue learning and growing.

I am so grateful I found seaweed, and this life.

Deep Breath

Being in the ocean for me is akin to being immersed in the Universal Spirit, or hanging out with God. It imparts a salty holiness in me each time I taste it. It is both serene and deadly, impossibly beautiful and unimaginably dark and barren.

The tides are a thing of mystery and magic, following the pull of the moon, but also the pull of the sun, every planet in our solar system and forces still unknown. The secrets that lie beneath the surface of the ocean are infinite. Indeed, the bottom of the ocean abyss has been called "the final frontier" after outer space because its pressure is so prohibitive. I have always dreamed about what lies beneath the sparkling waves and the still black water of night.

Because I loved the water so much, my parents put me on a swim team, but I could never figure out why anyone would spend all of their time swimming at the surface. Even in pools, where there was nothing to see, I would hover with my belly against the bottom, feeling the sensation of holding my breath under water.

When I was a little older, my sister and I would play a game in the pool where she would hold me up by one foot and I would have to try to get my head above the surface to take a breath. I remember being so desperate for air that I was sure I would die. I didn't. I learned early on that I could survive long past the moment when my mind told me I needed to breathe.

And so began a love affair with time spent under water.

The silent and eerie underwater world of the kelp forest clears away life's clutter, reminding me of my deep connection to and sense of wonder for the natural world.
CHRIS ADAIR PHOTO

Seaweed as Protector

The body is constantly bombarded with pathogens and environmental toxins that can harm it. This is why our bodies have an immune system: to fight off threats to our health. Incorporating seaweeds into your life is like having your own personal bodyguard everywhere you go. Seaweeds help the body to defend itself.

IMMUNE SUPPORT

The traditional use of seaweeds to enhance immunity is supported by current research, which shows that some seaweeds and seaweed extracts have an immunomodulatory effect (the ability to modify immune function). Of particular importance is fucoidan, a compound located in the cell walls of brown seaweeds, which studies show significantly increases immune function. Studies suggest that it does so in part by increasing macrophage phagocytosis (Itoh et al., 1993) and increasing T-cell interferon production (Maruyama et al., 2003).

TOXIN CLEANSING

When I teach people about seaweeds, I often say that if there are two words I want them to remember, they are "sodium alginate." Sodium alginate, the salt of alginic acid, is an amazing compound that safely cleanses toxins from the body. It is present in varying amounts in brown seaweeds. Kelp and seaweeds belonging to the Fucaceae family typically have the highest amounts of sodium alginate.

In the gut, sodium alginate binds to various environmental toxins, such as heavy metals, including mercury, lead, cadmium, strontium, aluminum and arsenic; radioactive isotopes; dioxins; and PCBs. They form an insoluble salt that does not dissolve in the bloodstream and is not broken down by saliva or bile. The salt, now containing the toxin, is safely excreted through the stool. Kelp purify the ocean in a similar way. They absorb and concentrate the minerals surrounding them, the good and the bad. The dangerous metals are harmful to seaweeds much like they are to us. Incredibly, seaweeds absorb the inorganic form of the metal, which is harmful to human health, and *convert* it to an organic form that is essentially biologically inert. This form passes through the body, without reacting or causing harm. Seaweeds are a powerful ally to help unload dangerous toxins from our bodies and prevent their accumulation in our tissues.

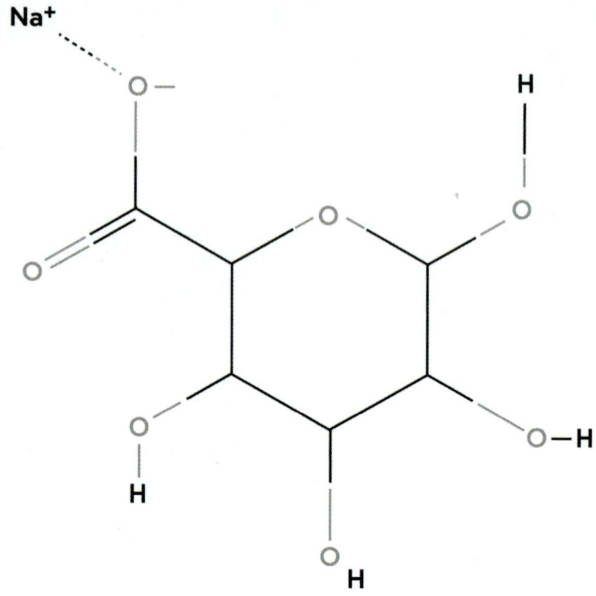

Sodium alginate.

RADIATION PROTECTION

Everyone is exposed to some radiation in daily life. There are two main mechanisms by which seaweeds offer protection. The first is that they are a rich source of natural iodine. Our thyroid gland, responsible for regulating metabolism as well as other essential functions, requires significant amounts of iodine to function properly. If it is saturated with natural iodine (I-127), it will not be able to absorb radioactive iodine (I-131). Radioactive iodine poses the most significant danger to human health after a nuclear fallout, and that is why governments recommend potassium iodide tablets when there is a threat of exposure. It is essential

Giant kelp (*Macrocystis pyrifera*).

that the thyroid is saturated with natural iodine *before* exposure if it is to be protective. Kelp are the richest source of iodine among seaweeds, and kombu (*Laminaria* spp.) contains the highest amounts. Furthermore, kelp store iodine in a form that is very bioavailable to humans, and they also contain complementary nutrients, such as selenium, that are essential to thyroid function.

Kelp and brown seaweeds that belong to the Fucaceae family, such as rockweed, offer a second form of protection against radioactive agents. The sodium alginate that is present in these species binds to some radioactive isotopes, including strontium-90, forming an insoluble salt that is safely excreted through the stool.

PROTECTION AGAINST PATHOGENIC BACTERIA

Seaweeds are exposed to many kinds of bacteria both in the ocean and, for intertidal seaweeds, in the air as well. All seaweeds contain antibacterial compounds that are uniquely evolved to protect them from pathogenic microbes. Hundreds of antibacterial compounds have been isolated from seaweeds and studied. Seaweeds and their extracts have shown antibacterial activity against both gram-positive bacteria (e.g. *Staphylococcus* and *Streptococcus*) and gram-negative bacteria (e.g. *E. coli*, *Salmonella* and *Listeria*).

Seaweeds have been added to animal fodder for centuries to improve growth and milk production and to prevent bacterial and parasitic infections. In a 2009 study by Wang et al., phlorotannins extracted from the brown seaweed *Ascophyllum nodosum* completely inhibited the growth of *E. coli* in vitro (on samples studied outside the body) over a twenty-four-hour incubation period. Studies on live animals also showed a decrease in the incidence of *E. coli* and other pathogenic bacteria (Jensen, 1972).

Additionally, phlorotannins extracted from brown seaweeds showed significant antimicrobial activity against foodborne pathogenic bacteria and methicillin-resistant *Staphylococcus aureus*, or MRSA (Eom, 2012; Choi et al., 2010).

PROTECTION AGAINST VIRUSES

Seaweeds contain potent and unique antiviral agents. Studies on antiviral compounds in seaweeds report a number of different mechanisms to fight viruses. These mechanisms include stopping some viruses from replicating, suppressing retroviral replication and preventing attachment of the virus to the host. Witvrouw and De Clercq (1997) reported that polysaccharides extracted from seaweeds inhibited the replication of the following strains of enveloped viruses (the class of virus that contains a membranous outer layer): herpes, HIV, flavivirus, arenavirus, orthopoxvirus, rhabdovirus, togavirus, HPV and influenza.

Carrageenan, extracted from red seaweeds, exhibits significant antiviral activity against HPV, both by preventing attachment of the virus to the host cell and through antiviral effects post-attachment (Buck et al., 2006). Several compounds isolated from red seaweeds have been reported to have specific antiviral activity against herpes viruses (Mahomoodally et al., 2019). Pharmaceutical companies have been funding research on red seaweeds and their antiherpes effect for years and several patents exist for this use.

PROTECTION AGAINST PATHOGENIC FUNGI AND YEASTS

Phlorotannins extracted from brown seaweeds have been shown to inhibit the growth of pathogenic fungi and yeasts, including specific activity against *Candida albicans*, a common cause of yeast infections (Lopes et al., 2013).

PROTECTION AGAINST PARASITES

Seaweeds have a long history of being used as a vermifuge (a treatment for parasitic worms) in many coastal areas, including Hawaii, New Zealand, Japan and Indonesia. Even today, many people use natural substances to remove intestinal worms and other parasites with effective results.

Kainic acid, extracted from the red seaweed *Digenea simplex*, is used as a traditional vermifuge in parts of Asia. Domoic acid, related to kainic acid, is extracted from *Chondria armata*, a red seaweed used as a traditional remedy for intestinal worms in Japan. Studies show that domoic acid has strong anthelmintic (worm-expelling) properties (Davis, 1998; Park et al., 1999). Extensive studies on chickens, pigs, cows and sheep showed a significant reduction in liver parasites when 3 to 5 per cent seaweed was added to their diet (Jensen, 1972).

Malaria, a disease caused by parasitic protozoa of the genus *Plasmodium*, kills hundreds of thousands of people each year, and according to the World Health Organization, children under the age of five are the most vulnerable. Malaria can be caused by a number of species of *Plasmodium*, but the majority of deaths are attributed to one species: *Plasmodium falciparum*. Seaweeds from all three groups (reds, greens and browns) have shown inhibitory activity against the harmful protozoa (Orhan et al., 2006; Topcu et al., 2003; Wright et al., 1996). Studies carried out on red seaweeds from the waters around France, for instance, showed strong, selective antiplasmodial activity against *Plasmodium falciparum*, with extracts from *Mastocarpus stellatus* having the strongest effect (Vonthron-Sénécheau et al., 2011). The brown seaweed *Undaria pinnatifida* (wakame), one of the richest known sources of fucoidan, showed inhibitory activity against strains of *Plasmodium falciparum* resistant to the malaria medication chloroquine (Chen et al., 2009).

SEAWEEDS AND ANTIOXIDANTS

Oxidation, a chemical process in the body, produces free radicals, which can damage cells and lead to an increased risk of chronic diseases and accelerated aging. Antioxidants are substances that inhibit oxidation and scavenge free radicals. In order to protect themselves from oxidative stress, seaweeds possess an abundance of powerful antioxidants.

Phlorotannins, which are found in highest amounts in brown seaweeds but are also present in smaller amounts in some green and red seaweeds, exert powerful free-radical scavenging. Zou et al. (2008) reported that phlorotannins extracted from the brown seaweed *Ishige okamurae* displayed more potent antioxidant activity than many commercially available medications with specific activity toward four harmful free radicals: hydroxyl, superoxide, alkyl and 1,1-diphenyl-2-picrylhydrazyl (DPPH). Phlorotannins decreased reactive oxygen species caused by UVB (Yan et al., 1997) and thus may be protective against sun damage. Dieckol, a phlorotannin extracted from the brown seaweed *Ecklonia cava*, is a particularly potent antioxidant.

Fucoxanthin, a carotenoid pigment present in brown seaweeds, and fucoidan, a sulphated polysaccharide found in brown seaweeds, exhibit strong antioxidant activity. Ascorbic acid (vitamin C), present in edible seaweeds, is also a powerful scavenger of free radicals.

ANTI-INFLAMMATORY PROPERTIES OF SEAWEEDS

Chronic inflammation has been linked to the development of many diseases. Seaweeds have long been used in traditional medicine to treat disorders associated with inflammation. The edible red seaweed *Pyropia dentata*, for instance, was used in various regions to treat bronchitis, enlarged lymph nodes and hypersensitivity (Kazłowska, 2010). Current research recognizes seaweeds as an excellent source of anti-inflammatory agents and has identified anti-inflammatory mechanisms, including oxygen free-radical scavenging and the inhibition of inflammation-mediating enzymes. Seaweeds from all three groups (browns, reds and greens) have shown anti-inflammatory activity in vitro and in vivo (within living specimens), with phlorotannins, fucoxanthin and fucoidan exhibiting the most significant anti-inflammatory activity.

SEAWEEDS AND ALLERGIES

Seaweeds are used in Korean folk medicine to treat allergies, and studies on seaweeds from all three groups (reds, greens and browns) have reported antiallergic effects both in vitro and in vivo. Compounds extracted from brown seaweeds (including phlorotannins, alginic acid and fucoidan) and red seaweeds (including porphyrin) were shown to significantly inhibit allergic response (Le et al., 2009; Li et al., 2008; Maruyama et al., 2005; Vo, Ngo and Kim, 2012; Yoshida et al., 2004) by regulating immune response and inhibiting the release of histamine. Regulating immune response is considered to be essential in the treatment of chronic allergies, while histamine causes many of the undesirable symptoms.

NEUROPROTECTIVE PROPERTIES OF SEAWEEDS

The Japanese population experiences low rates of dementia (Ozawa et al., 2013) and Parkinson's disease (Okubo et al., 2012), which some research suggests could be linked to the regular consumption of seaweed. Over the last decade, intensive studies have uncovered a variety of neuroprotective compounds contained in seaweeds and the mechanisms they use to protect the brain. Additionally, alginates from kelp and some other brown seaweeds safely cleanse harmful, neurotoxic heavy metals, such as mercury and aluminum, helping to prevent their accumulation in the brain. Seaweeds are rich in minerals essential for healthy brain function, including rare trace minerals such as selenium and chromium, and nutrients such as choline.

Alzheimer's disease, Parkinson's disease and Huntington disease are all neurodegenerative disorders associated with the presence of specific amyloids, and a number of studies indicate the potential for seaweeds to reduce the formation of these pathogenic amyloids. Seaweeds living in extreme conditions are of particular interest due to their incredible ability to protect their own proteins from misfolding (a process that takes place in the course of neurodegenerative diseases). For instance, extracts of *Alaria esculenta*, a brown seaweed that thrives in the harsh environment of the Bay of Fundy in eastern Canada, were found to reduce the formation of α-synuclein amyloids, which are associated with Parkinson's disease (Giffin et al., 2017).

Alzheimer's disease, the most common neurodegenerative disorder, is a horrible, fatal disease that literally robs people of their memories. It is associated with the formation of β-amyloids in the brain. Reducing these amyloids, especially

Medicinal bounty from the sea.

in the early stages of Alzheimer's, is a key target for treatment; however, no known treatment is curative. Fucoidan, extracted from brown seaweeds, exhibits a neuroprotective effect on the activity of β-amyloids and reduced neuron death (Gao et al., 2012).

Alzheimer's is also associated with a deficiency of the neurotransmitter acetylcholine, and most available drugs for Alzheimer's patients inhibit acetylcholinesterase (the enzyme responsible for breaking down acetylcholine). Phlorotannins extracted from brown seaweeds have been shown to have an inhibitory effect on acetylcholinesterase (Yoon, Ham et al., 2009). Fucoidan, extracted from brown seaweeds, improved learning and reduced memory impairment induced by β-amyloids in animal models. It does this, the authors of a 2012 study propose, through the following mechanisms: regulating the cholinergic system, reducing oxidative stress and inhibiting cell apoptosis (Gao et al., 2012).

Neuroinflammation is also positively associated with Alzheimer's disease. Seaweeds and seaweed extracts—including fucoidan, from brown seaweeds (Cui et al., 2010); dieckol, a phlorotannin extracted from the brown seaweed *Ecklonia cava* (Jung et al., 2009); and the green seaweed *Ulva conglobata* (Jin et al., 2006)—have shown inhibitory effects against several neuro-proinflammatory enzymes.

IN SUMMARY

Seaweeds are a centuries-old medicinal food traditionally used to enhance one's resilience and vitality and provide protection from environmental pathogens and mental deterioration. Current research validates these traditional uses and suggests that incorporating seaweeds into one's diet can support overall health.

Fucoidan

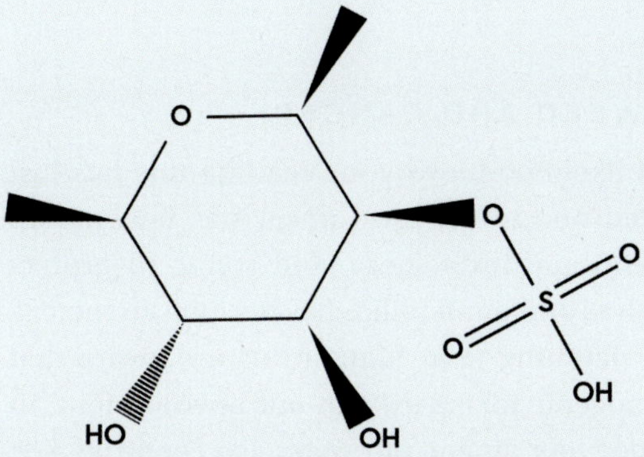

Fucoidan.

Located primarily in the cell walls of brown seaweeds, fucoidans provide structure and are thought to help protect the seaweed against desiccation (drying). Fucoidan was first isolated from the brown seaweeds *Ascophyllum nodosum*, *Fucus vesiculosus*, *Laminaria digitata* and *Laminaria saccharina* (now *Saccharina latissima*) by Professor Harald Kylin of Uppsala University, Sweden, in 1913. Until recently, fucoidan was thought to be unique to brown seaweeds, but it has since been discovered in a few species of marine invertebrates whose diet includes seaweeds, including sea cucumbers, and in the egg coating of sea urchins.

Fucoidan, a water-soluble dietary fibre, is the subject of well over a thousand studies. Currently, it is being exhaustively investigated for its potential role in cancer prevention and therapy, and it has been shown to help alleviate undesirable side effects of chemotherapy and radiation. Fucoidan extracts from brown seaweeds are sold and marketed as a nutraceutical for the prevention of cancer and as a supportive therapy.

Fucus spiralis.

Research has also shown fucoidan to have antiviral, antibacterial, anti-inflammatory, anticoagulant, antiadhesive, anti-obesity, cardioprotective, neuroprotective and immunomodulatory effects. Furthermore, studies have shown fucoidan to help normalize blood sugar levels, and suggest potential as an osteoporosis preventative and allergy treatment. Studies on fucoidan have also reported numerous positive effects on gastrointestinal health.

Fucoidan is bioavailable simply by consuming most brown seaweeds. It is safe and non-toxic and can be consumed by people of all ages. In my research, I found more positive attributes associated with fucoidan than any other novel and unique compound in seaweeds, and it appears to have the most diverse potential for supporting human health. Phycologists have studied its role in protecting seaweed; perhaps it protects us in a similar way.

Seaweed and Chronic Diseases

SEAWEED AND CANCER

In all likelihood, everyone reading this has lost a loved one to cancer. Cancer has become an epidemic in our society, and trying to protect ourselves and our families has become an increasingly daunting task. Many of us are aware that carcinogenic toxins exist in our environment: in our food, our air and our water. For centuries, seaweeds have been used in the treatment of cancer. In the ancient Egyptian medical text the Ebers papyrus, seaweed is prescribed as a treatment for breast cancer. Traditional Chinese Medicine and Japanese folk medicine include the use of specific seaweeds for the treatment of tumours.

In addition to the immune function support, abundant nutrients, antioxidants and toxin cleansing properties that seaweeds possess, some of their unique biologically active compounds show further potential in disrupting some of the key processes that lead to cancer.

The following compounds unique to marine algae have been studied for their potential benefit in the prevention and/or treatment of cancer: fucoidan, fucoxanthin, phlorotannins, carrageenan, laminarin, alginates, porphyran and bromoditerpenes. Of these compounds, fucoidan is currently the most extensively studied and shows the most potential benefit at this time. The majority of these compounds exist in the brown seaweeds.

One of the primary ways that cancer is able to progress in the body is through the deactivation of cell apoptosis (programmed cell death): the body's immune system fails to locate, identify and destroy the invasive cancer cells. Indeed, inactivation of cell apoptosis is considered to be one of the foundational markers of cancer, and activating cell apoptosis is one of the goals of treatment (Brown and Attardi, 2005). Fucoidan, present in many brown seaweeds, was shown to induce cell apoptosis on human cancer cells in vitro, including lymphoma (Aisa et al., 2005), gastric adenocarcinoma (Park, Kim et al., 2011), pancreatic cancers (Delma et al., 2015), breast cancers (Zhang et al., 2013) and various colon cancers (Kim, Park et al., 2010).

Studies suggest fucoidan impairs tumour progression and has a role in controlling key processes of tumour genesis. Specific studies report the inhibition of cell proliferation on bladder cancer cells treated with fucoidan (Cho et al., 2014), antiangiogenic (inhibiting the growth of new blood vessels) activity on uterine carcinoma cells treated with fucoidan (Ye et al., 2005), and an antimetastatic effect on lung cancer cells treated with fucoidan (Lee, Kim and Kim, 2012; Huang et al., 2015).

Fucoidan also increases immune function, specifically increasing macrophage phagocytosis and the production of T-cell interferons and interleukins. It also shows excellent selective cytotoxicity, meaning that it selectively kills cancer cells while leaving healthy cells alone. Fucoidan is also useful for alleviating the side effects following chemotherapy and radiation treatment and restoring vitality, including immune support, to the body.

More than a thousand scientific studies have been done on fucoidan, and researchers report anticancer activity, often in a dose-dependent manner, on a variety of different cancer types, including lymphoma, leukemia, and gastric, colon, breast, lung, bladder, uterine, pancreatic and prostate cancers, with no toxicity observed. The limitations of this research are that it is mostly on isolated human cancer cells in vitro or in animal models in vivo. More studies on human subjects are indicated.

Fucoxanthin, a carotenoid pigment that gives brown seaweeds their yellowish-brown colour, has also been studied for its potential therapeutic benefits. Research has identified the following anticancer mechanisms associated with fucoxanthin: antiproliferation, induction of cell apoptosis, cell cycle arrest and antiangiogenesis.

Fucus species such as rockweed are rich in fucoidan.

Research reporting an antiproliferation effect of fucoxanthin include studies on prostate cancer (Kotake-Nara et al., 2001), leukemia (Kim, Heo, Kang et al., 2010), colon cancers (Hosokawa et al., 2004), melanoma (Kim et al., 2013), liver cancer (Liu et al., 2009) and gastric adenocarcinoma (Yu et al., 2011). Research reporting fucoxanthin's ability to induce apoptosis includes studies on leukemia, breast and colon cancer (Konishi et al., 2006), lymphoma (Yamamoto et al., 2011) and bladder cancer (Zhang et al., 2008). Research reporting a positive effect of fucoxanthin on cell cycle arrest includes studies on liver cancer (Liu et al., 2009) and sarcoma (Wang et al., 2012). Research suggesting an antiangiogenic effect of fucoxanthin includes studies by Sugawara et al. (2006) and Ganesan et al. (2013). In a 2008 study by Zhang et al., the authors reported that apoptosis on human bladder cancer cells in vitro rose 93 per cent within seventy-two hours of being treated with fucoxanthin. Researchers report that anticancer effects are dose-dependent (their effects increase with increased administration of fucoxanthin), and that induction of cell apoptosis by fucoxanthin is selective, preferentially targeting cancer cells and leaving healthy cells alone. In addition, a diet that includes carotenoid-rich foods has been linked to a reduced risk of lung cancer (Holick et al., 2002). Again, this research is mostly from in vitro studies on isolated human cancer cells or in vivo animal models.

Phlorotannins are phenolic compounds that exist in the highest amounts in brown seaweeds but are also found in smaller amounts in some red and green seaweeds. Extracts from the brown seaweeds *Nereocystis luetkeana* (bull kelp), *Macrocystis integrifolia* (giant kelp) and *Laminaria setchellii* (kombu), and from the red seaweed *Palmaria palmata* (dulse), showed antiproliferative activity on cervical cancer cells linked to the high phlorotannin content (Yuan and Walsh, 2006), while phlorotannins extracted from the brown seaweed *Sargassum muticum* displayed antiproliferative activity on breast cancer cells (Namvar et al., 2013). Studies on phlorotannins extracted from the brown seaweed *Ecklonia cava* have reported the following: antiproliferative activity on animal models in vivo against colon cancer and melanoma and in vitro against human leukemia cells (Athukorala et al., 2006), inhibition of metastasis in fibrosarcoma (Zhang et al., 2011), inhibition of angiogenesis and inhibition of the activity of matrix metalloproteinases, enzymes involved in the cancer process (Kim et al., 2006). Phlorotannins extracted from the red seaweed *Gracilaria corticata* and green seaweeds from the genus *Enteromorpha* showed antioxidant activity correlated with an antiproliferative effect on liver and breast cancer cells (Narasimhan et al., 2013). The phlorotannin phloroglucinol extracted from brown seaweeds may help to make anticancer drugs (including Taxol, cisplatin and etoposide) more effective by increasing sensitivity of breast cancer cells to the drugs and through inhibited proliferation of breast cancer cells, targeting cancer stem-like cells (CSCs), which are proposed to be responsible for relapse after cancer treatment (Kim, Suh et al., 2015). The strong antioxidant activity of phlorotannins is described as a primary driver of their anticancer activity and also helps protect cells from injury caused by radiation treatments (Shin et al., 2014).

Porphyran, a sulphated carbohydrate and powerful antioxidant from the red seaweed genus *Pyropia* (formerly *Porphyra*), has reported antitumour activity. Studies support its induction of selective cell apoptosis on gastric cancer cells (Kwon and Nam, 2006).

The relationship between dietary seaweed and cancer incidence has also been studied. In animal model studies in vivo, Teas et al. (1984) found that adding 5 per cent of the kelp *Laminaria angustata* to one's diet was effective at slowing the development of breast tumours. Yamamoto et al. (1986) reported that after five weeks of administration of kelp from the genus *Laminaria*, inhibition of the growth of sarcoma cells ranged from 70.3 to 83.6 per cent with dietary seaweed and 61.9 to 95.2 per cent with intraperitoneal injections (injections delivered into the peritoneum, or body cavity).

In Japan, where seaweed consumption is among the highest of any population in the world, cancer rates are among the lowest in the developed world, yet have been increasing since 1945. Oiso (1975) and Wynder and Hirayama (1977) propose that this increase is caused by the progressive Westernization of the Japanese diet. A Western diet is one high in saturated fats and sugar and low in fibre. Furthermore, first- and second-generation Japanese migrants to the United States had increased rates of breast cancer and colon cancer compared with the Japanese population (Weisburger et al., 1980). Regular dietary intake of seaweed, specifically the consumption of wakame (*Undaria pinnatifida*) and kombu (*Laminaria* spp.), both kelp, has been theorized to be responsible for the much lower rates of breast cancer and breast cancer–related mortality in postmenopausal Japanese women. The results of research on human subjects support the hypothesis that dietary intake of seaweed has a preventative effect on the incidence of cancer. This effect is illustrated by a clinical study on breast cancer by Dr. Jane Teas and colleagues in 2013 and an investigation into diet and colorectal cancer that found seaweed consumption to be inversely related to colon and rectal cancer in a dose-dependent manner (Hoshiyama et al., 1993). Stimulation of immune response, presence of non-digestible fibre and antibiotic substances that may influence fecal ecology are among the mechanisms proposed by Teas (1983) for the relationship between dietary seaweed consumption and lower cancer incidence.

Macrocystis pyrifera (giant kelp) with sea anemone.

Additionally, studies dating back to 1924 identify increased cancer incidence in populations with iodine deficiency. Low iodine is theorized to increase the incidence of cancers such as breast and stomach cancer. The anti-inflammatory, antioxidant and antiproliferative effects of iodine are proposed as some of its anticancer mechanisms. Seaweeds are a rich source of natural iodine, with kelp containing the highest amounts.

Most studies on the chemotherapeutic properties of seaweeds in relation to cancer are on brown seaweeds, specifically kelp (seaweeds belonging to the order Laminariales) and seaweeds belonging to the order Fucales. All of these species contain fucoidan, fucoxanthin and sodium alginate and are rich in phlorotannins. Brown seaweeds also contain phytoestrogens, which reduce estrogen in the body and therefore may be helpful in preventing estrogen-dependent cancers such as breast cancer.

Many studies on seaweeds have highlighted the potential cancer preventative and therapeutic effects of whole seaweeds and their extracts without isolating specific compounds. Furthermore, studies designed to determine the safety of isolated compounds including fucoidan, fucoxanthin and phlorotannins have reported all three to be non-toxic (Kim, Lee et al., 2010; Beppu et al., 2009; Hwang, Terada and Shin, 2008).

SEAWEED AND CARDIOVASCULAR DISEASE

According to the World Health Organization, CVD is the leading cause of death worldwide. It includes an array of conditions that involve the heart muscle and blood vessels, including coronary artery disease, congestive heart failure, venous thrombosis and others. A healthy cardiovascular system is one in which the heart muscle is strong, with good contractual force, and the blood vessels are elastic, clean and toned. Seaweeds are abundant in vitamins and minerals that nourish and tone the heart and blood vessels, and they possess unique compounds that research suggests are cardioprotective. The brown seaweed kombu (*Saccharina* and *Laminaria* spp.), for instance, is highly regarded as a cleanser and tonic to the entire cardiovascular system and has been used for centuries in Japan to treat hypertension and other cardiovascular issues. Regular consumption of seaweed has been associated with a lower incidence of CVD risk factors, including hypertension, hyperlipidemia, diabetes, obesity and metabolic syndrome (Baik et al., 2013; Cardoso et al., 2015).

Seaweeds contain many nutrients essential for healthy blood vessels and a healthy heart. For example, they are a rich source of both soluble and insoluble fibre. Fibre helps lower the presence of low-density lipoproteins (LDLs) in the blood, the accumulation of which can lead to cholesterol buildup in the arteries and atherosclerosis. For this reason, LDLs are sometimes referred to as "bad" cholesterol. High-density lipoproteins (HDLs), on the other hand, are often referred to as "good" cholesterol because they carry cholesterol back to the liver to be metabolized. Studies show that low levels of HDLs can increase the risk of heart disease. Seaweeds and their extracts have been shown to increase HDL levels and lower LDL levels (Bocanegra et al., 2009; Kandale et al., 2011; Pengzhan et al., 2003).

Vitamins, found in abundance in seaweeds, also contribute to a healthy heart, healthy vessels and healthy blood. Some of the essential vitamins that are required for healthy cardiovascular function—but which must be obtained from external sources—include vitamin B1 (thiamine), which is essential for healthy nerve and muscle formation and heart function; vitamin B2 (riboflavin), which assists in the absorption of iron and the development of healthy blood cells; and vitamin K, an essential component of healthy blood that also helps to regulate blood calcium.

Seaweeds are a bioavailable and abundant source of minerals essential to healthy blood, including copper, iron, sodium, potassium, calcium and magnesium. An ideal ratio of potassium to sodium in the blood is critical to health and affects every part of the cardiovascular system. The salts in seaweeds are potassium-rich and help to regulate blood pressure and prevent vascular calcification (mineral deposits in the arteries that can contribute to atherosclerosis). Hypertension is associated with a deficiency in calcium, magnesium and potassium. Seaweeds such as winged kelp (*Alaria marginata*), which are rich in magnesium, iron and potassium and are among the highest known sources of calcium, nourish the body with minerals required for the healthy function of the cardiovascular system. Anemia is a

Laminaria setchellii, the Pacific Northwest's kombu.

condition characterized by deficient red blood cells, resulting in less oxygen in the blood, which forces the heart to work harder. This disease is often caused by a deficiency in or difficulty absorbing minerals such as iron and vitamins such as B12. Individuals suffering from anemia can benefit from the abundant bioavailable minerals and vitamins in seaweeds, which include iron and vitamin B12.

The anti-inflammatory properties, antioxidants and gastrointestinal support from seaweeds also contribute to cardiovascular health. Fucoidan, a compound present in brown seaweeds, has been extensively studied for its cardioprotective properties. Research on fucoidan has reported the following: reduced cholesterol and LDLs and increased HDLs (Huang et al., 2010); anti-atherosclerotic activity, significantly inhibiting arterial plaque formation (Wang et al., 2016); reduced formation of atherosclerotic lesions in the aorta (Park et al., 2016); reduced plaque, cholesterol, tissue weight (liver and white adipose) and blood lipids (Yokota et al., 2016); and antithrombotic activity in human subjects (Ren et al., 2013). Furthermore, fucoidan exhibits an anticoagulant effect and has been shown to significantly reduce hypertension (Cui et al., 2014).

Alginates, extracted from brown seaweeds in the orders Laminariales (kelp) and Fucales, also exhibit cardioprotective activity. Studies show that alginates help lower cholesterol, reducing LDLs and increasing HDLs (Kartika, Bangun and Rosidah, 2018). In a study on human subjects by Paxman et al. (2008b), sodium alginate from brown seaweeds was found to decrease risk factors for CVD by reducing the absorption of cholesterol and glucose. Alginates bind to sodium in the intestines and remove it from the body via the stool, helping to reduce sodium in the bloodstream, an excess of which can lead to high blood pressure.

Laminarin, a compound found in some brown seaweeds and in the highest amounts (up to 50 per cent dry weight) in kelp and in seaweeds belonging to the order Fucales, has anticoagulant properties and has been reported to lower blood pressure and inhibit arterial plaque and blood lipids, which can lead to atherosclerosis (Patil et al., 2018).

Studies on fucoxanthin, a carotenoid found in brown seaweeds, suggest significant cardioprotective properties and report the following benefits: anti-obesity effects and improved lipid metabolism (Muradian et al., 2015); delayed development of stroke signs and protection against neuronal cell death after stroke (Ikeda et al.,

2003); and improved liver and plasma lipid levels and lowered blood glucose (Woo et al., 2010). Fucosterol, a compound extracted from brown seaweeds, has also been shown to lower blood pressure (Hagiwara et al., 1986).

While most studies on the cardioprotective properties of seaweeds are in vitro or in vivo animal model studies, there have been clinical trials on human subjects and some etiological research.

The Japanese population, whose diet is typically high in fish, soybean products, fruits, vegetables and seaweeds and low in red meat and dairy, have low mortality rates due to CVD compared with other developed nations. In a study on 40,547 Japanese men and women, results showed that the traditional Japanese diet resulted in a decreased risk of developing CVD (Shimazu et al., 2007). In a study of the Okinawan people of the Okinawa Islands in Japan, a population known for its exceptional longevity, the inclusion of fish, soy, green vegetables and seaweed in the diet was identified as a potential source of chronic non-communicable disease prevention (Yamori, Miura and Taira, 2001). A study on hypertensive human subjects reported a reduction in blood pressure and cholesterol with the consumption of an average of 3.3 grams per day of the brown kelp *Undaria pinnatifida* (wakame) over an eight-week period (Hata et al., 2001).

Seaweeds provide nutrients essential to the health of the heart, blood and blood vessels. The traditional use of seaweeds to regulate, cleanse and tone the cardiovascular system is supported by current research, which describes significant cardioprotective properties. Just a small amount of seaweed eaten consistently, especially kelp, can contribute to a heart-healthy diet.

SEAWEED AND DIABETES

Diabetes mellitus is a serious chronic disease in which either the pancreas does not produce enough insulin or the body cannot effectively utilize insulin. It is characterized by high levels of glucose in the blood, which can cause serious damage to the body's systems, especially the nerves and blood vessels, and can lead to CVD, kidney failure and other maladies. Type 2 diabetes can be largely controlled by diet and exercise. The number of people suffering from diabetes is increasing globally. Research describes therapeutic benefits of seaweeds and their extracts for the prevention and management of diabetes.

Alaria marginata (winged kelp) with blood star.

Fibre is the largest component of the biomass of seaweeds: they are a rich source of both soluble and insoluble fibre. Diets high in fibre are associated with a decreased risk of developing type 2 diabetes. Soluble fibre slows the digestion of food, which in turn slows the absorption rate of

Dried Seaweed

Drying seaweeds is the best way to preserve them, and dried seaweeds are essentially a living food. Remember that seaweeds do not have roots, so they don't die when they are cut. When dried seaweeds are reconstituted in water, it is as if the tide has come in and they come back to life with vitality and vigour.

Heart-shaped *Pyropia abbottiae*, black nori, dried on a screen.

glucose into the bloodstream following a meal. A diet low in fibre causes digestion to be fast and a sudden, large amount of glucose to be absorbed into the blood with each meal. This stresses the pancreas to produce large amounts of insulin all at once, and eventually the pancreas cannot keep up enough insulin production, leading to type 2 diabetes. Agar from red seaweeds and fucoidan and alginates from brown seaweeds are examples of soluble fibres that help lower blood glucose levels. Additionally, seaweeds are an abundant, bioavailable source of minerals, including the trace minerals chromium and vanadium, which are associated with balancing blood sugar.

Seaweeds are a rich and varied source of antioxidants, which help lower blood glucose levels. Phlorotannins, found in the highest quantity in brown seaweeds but also present in some red and green seaweeds, are powerful antioxidants that have shown antidiabetic effects in numerous studies. Dieckol, a phlorotannin extracted from the brown seaweed *Ecklonia cava*, has been shown to significantly decrease blood glucose levels. It showed strong inhibitory activity against α-glucosidase and α-amylase, enzymes that metabolize carbohydrates into glucose and that are responsible for sudden increases in blood glucose levels, causing hyperglycemia in type 2 diabetics. Inhibiting α-glucosidase and α-amylase is considered an important strategy in managing type 2 diabetes. Research has found that the inhibitory action of dieckol was stronger than the commercial drug acarbose and showed no toxicity (Lee et al., 2010). Lee, Min et al. (2012) reported that dieckol also reduced hepatic glucose production, improved glucose and lipid metabolism and significantly lowered blood glucose levels. Furthermore, their results suggest that dieckol may improve insulin sensitivity. Studies also showed that dieckol protected cells against hyperglycemia-induced oxidative stress (Lee et al., 2010).

In addition, Mochizuki et al. (1995) observed a dose-dependent hypoglycemic effect of *Ecklonia cava*, which they attributed to the delay in gastric emptying due to fibre content. In a study by Kang et al. (2010), a methanol extract of *Ecklonia cava* was found to significantly reduce blood glucose and increase insulin concentration in type 1 diabetic rats, as well as improving symptoms of diabetes including polydipsia (excessive thirst) and polyphagia (excessive hunger). Extracts of the brown seaweed and kelp *Laminaria japonica* reduced blood glucose levels through what researchers believe was antioxidant activity (Jin et al., 2004). Butyl isobutyl phthalate, a bioactive compound isolated from *Laminaria japonica*, significantly reduced blood sugar in vivo (Bu et al., 2010).

Fucoidan, a compound found in brown seaweeds, significantly prevented hyperglycemia and reduced blood glucose in vivo (Kim et al., 2012), and fucoidan extracted from *Saccharina japonica* decreased blood glucose and increased serum insulin in vivo (Wang et al., 2013). Researchers suggest that stimulating the pancreas to release insulin and reducing insulin metabolism are possible mechanisms for this significant hypoglycemic effect.

In a study on human subjects, Paradis et al. (2011) found that a single ingestion of 0.5 grams of a mixture of the brown seaweeds *Fucus vesiculosus* and *Ascophyllum nodosum* after a carbohydrate-rich meal increased insulin sensitivity. Kim et al. (2008) reported a significant reduction in fasting glucose and blood glucose levels two hours postprandial (after a meal) in type 2 diabetic human subjects ingesting 48 grams per day of the brown seaweeds kombu (*Laminaria japonica*) and wakame (*Undaria pinnatifida*) after four weeks. Furthermore, dietary fibre was 2.5 times higher in the group ingesting the seaweed and HDL levels also increased significantly.

Seaweeds show therapeutic benefits for diabetes specifically by helping to balance blood glucose levels.

IN SUMMARY

Both traditional medicinal uses of seaweed and current scientific research illustrate the extensive and impactful benefits of seaweed in the protection against chronic illnesses such as cancer, CVD and diabetes. Including a small amount of seaweed in your regular diet is an excellent way to support your overall health. If you are new to cooking with seaweeds, sprinkling one-half to one teaspoon of dried seaweed flakes over top of a meal is an excellent way to begin incorporating seaweeds into your diet. In addition to the recipes in this book (see page 184), Prannie Rhatigan's *The Irish Seaweed Kitchen* gives excellent instructions for people who want to start cooking with seaweeds.

JENNIFER JELLETT PHOTO

Green Soul

The life force in the rainforest is so dense you can almost touch it with your hands. There is a plethora of green in every shade and evidence of moisture in every fold.

The perfectly straight trunk of the Sitka spruce has puzzle-piece bark stretching up its formidable length and stands guard over the shore. The open ocean swells become waves, crashing on the northwestern shores, slowly turning cliffs to rocks, rocks to pebbles and pebbles to sand. The mist clings to jutting trees or points of land and the clouds are so low it seems you could reach them simply by stretching your arms above your head.

The mist through the trees colours the air a shade of slate grey. The sound of the rain is soothing and is sometimes pierced by the sad cry of a gull. The trees are the only part of the vista that isn't a shade of grey, and they glow shades of brilliant emerald green through the mist.

I am one of those blessed to call the Pacific Northwest my home. Cascadia. Land of the misty temperate rainforest, the rarest and most endangered rainforest type in the world. I immerse myself in its vitality and thickness of life, which it exhales into my inhale, and I breathe it deep into my lungs and lifeblood.

I am one of those blessed to call the Pacific Northwest my home. Cascadia. Land of the misty temperate rainforest, the rarest and most endangered rainforest type in the world.

Fucoxanthin

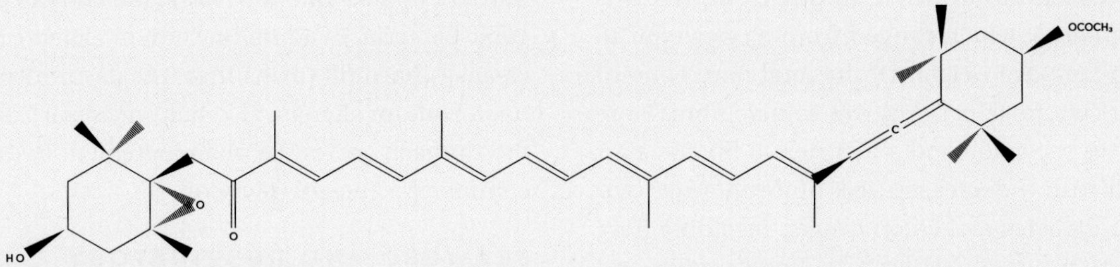

Fucoxanthin.

Fucoxanthin is a carotenoid pigment found in brown seaweeds, used to absorb light energy for photosynthesis and to provide photoprotection. All brown seaweeds contain fucoxanthin as their dominant pigment, which gives them their golden-brown colour. Fucoxanthin is also present in diatoms, a group of brown microalgae. Scientists Richard Willstätter and H. Page first isolated fucoxanthin in 1914 from the brown seaweeds *Fucus* spp., *Laminaria* spp. and *Dictyota* spp.

There have been many papers published on fucoxanthin and its potential role in promoting human health. Studies have reported anticancer effects including induction of cell apoptosis, antiproliferation, cell cycle arrest and antiangiogenesis on cancer cell lines including prostate, breast, colon, liver, bladder, stomach, melanoma, leukemia and lymphoma. Studies report the cytotoxic effect of fucoxanthin to be selective, meaning it targets cancer cells while leaving healthy cells alone. Studies on the inhibitory effect of fucoxanthin on prostate cancer cells have been particularly encouraging.

Additionally, studies on fucoxanthin have reported antidiabetic, anti-inflammatory, antioxidant, cardioprotective, neuroprotective and osteoprotective activity.

Seaweed and Body System Support

SEAWEED AND A HEALTHY GUT

Seaweeds have been used traditionally to relieve constipation, soothe stomach pain associated with indigestion and alleviate flatulence. Cooking beans with seaweed is a time-honoured way of reducing flatulence associated with legume digestion. Gastrointestinal health, in particular healthy gut microbiota, is now known to be linked to one's overall health. An estimated 70 per cent of our immune system resides in the gut. Seaweeds support gut health in a number of ways: they are high in fibre and a rich source of bioavailable vitamins and minerals, they cleanse toxins from the gastrointestinal tract, they provide barrier protection to the lining of the gut and mounting evidence strongly suggests they are a source of prebiotics that can favourably modulate gut microbiota.

Seaweeds are an abundant source of fibre and contain both soluble and insoluble fibre. Lahaye (1991) determined the fibre content of seven edible seaweeds from all three groups (reds, browns and greens), which ranged from 33 per cent to 75 per cent dry weight. Of the total fibre content, 52 per cent to 85 per cent was soluble fibre. Fibre cleanses the colon and a diet rich in fibre is associated with a decreased risk of developing CVD, type 2 diabetes and colon cancer. Insoluble fibre helps to promote regular bowel movements and healthy stool formation, and soluble fibre aids in weight loss, lowers cholesterol and helps to regulate blood sugar by slowing the release of glucose into the bloodstream. Examples of seaweed fibres are fucoidan and alginate from brown seaweeds, carrageenan and agar from red seaweeds and ulvan from green seaweeds.

Seaweeds contain abundant vitamins and bioavailable minerals essential to gut health, including vitamin B-complex, vitamin C, vitamin D, iron, magnesium, calcium, zinc and selenium. A mineral extract from red seaweeds containing calcium, magnesium, phosphorus, selenium and other trace minerals reduced the development and severity of mild to moderate colitis in an in vivo animal model study (Aviello et al., 2014).

Prebiotics nourish beneficial bacteria in the gut, improving gut microbiota, which in turn provides systemic benefits to the host. There is strong evidence that polysaccharides in seaweeds, including alginate, fucoidan, laminarin, agar and carrageenan, are sources of prebiotics (O'Sullivan et al., 2010; Ramnani et al., 2012). In a study on the microbiota of pigs, fucoidan extracted from brown seaweeds increased the concentration of the beneficial bacteria *Lactobacillus* spp., and the polysaccharide laminarin, extracted from brown seaweeds, reduced the concentration of the pathogenic bacteria *Enterobacter* spp. (Lynch et al., 2010).

Studies also suggest that sodium alginate, found in brown seaweeds, improves the barrier quality of the mucosal lining of the gut by decreasing permeability (Mackie et al., 2016). A healthy gut lining is necessary for regulating water and nutrient uptake and preventing the entry of pathogenic bacteria. In addition, sodium alginate safely cleanses harmful toxins from the gastrointestinal tract. Sodium alginate also helps to absorb acid in the stomach and is the main ingredient in the the commercial antacid Gaviscon.

SEAWEED AND RESPIRATORY SUPPORT

Seaweeds have a long and rich history of being used in myriad ways to support the respiratory system. For example, some red seaweed species that are rich in carrageenan have been used for millennia to treat respiratory ailments such as sinus infections and pneumonia (Kandale et al., 2011). In Ireland, the carrageenan-rich red seaweeds *Chondrus crispus* and *Mastocarpus stellatus* are traditionally used to treat chest problems such as pneumonia, bronchitis, sinus infections, coughs and sore throats and are valued as an effective expectorant to expel phlegm. Thalassotherapy (bathing in seaweeds), infusions, steam inhalation, poultices and the consumption of seaweeds are all methods used to heal afflictions of the respiratory system.

The antiviral, antibacterial and antifungal components of seaweeds help fight infections, and the mucilaginous (sticky, mucus-like) components are soothing and healing to mucous membranes. The immune-enhancing and anti-inflammatory activity of seaweeds also aids in healing and supporting the respiratory system. Furthermore, seaweeds are a rich source of omega-3 fatty acids, which studies have shown are helpful in reducing the symptoms of asthma. In addition, some seaweeds contain high amounts of desirable fatty acids, also found in cold-water fish, such as eicosapentaenoic acid and docosahexaenoic acid (van Ginneken et al., 2011). One epidemiological study examined the effect of the dietary intake of

Red seaweeds such as *Chondracanthus exasperatus* (Turkish towel) are rich in carrageenan.

fish, fatty acids and seaweed on the incidence of asthma in a Korean population aged nineteen to sixty-four; researchers found that subjects with medication-prescribed asthma ate significantly less seaweed (Kim and Ju, 2019).

There is also some evidence of protective properties of seaweeds against cigarette smoke toxicity. In a study by Jang et al. (2016), fucoxanthin extracted from a brown seaweed showed significant and multi-faceted protection from cigarette smoke-induced damage to human bronchial cells in vitro.

SEAWEED AND BONE HEALTH

Seaweeds contain an abundance of minerals that are essential to bone health, including calcium, magnesium, phosphorus, potassium and trace minerals including iron, copper, boron, silicon, fluoride and manganese. These minerals are stored in chelated and colloidal forms, making them very bioavailable. The richness and diversity of minerals in seaweed cannot be overstated. Up to 39 per cent of the dry weight of edible seaweeds is minerals (Rupérez, 2002). Seaweeds also contain an abundance of vitamins required for healthy bones, including vitamins A, B-complex, C, D and K.

Studies suggest that seaweeds and their extracts significantly increase bone calcium content (Yamaguchi et al., 2001). Seaweeds contain very high amounts of calcium. Most edible seaweeds contain approximately six times more calcium than cow's milk, some containing levels much higher than that. Seaweeds also contain complementary nutrients required for the optimal absorption of calcium, such as vitamin D and other minerals. A combination of minerals isolated from a *Lithothamnion* species (a red seaweed) reduced bone mineral loss, increased bone mineral density and improved bone strength in a study on female mice. The authors suggest that the trace minerals in the seaweed-derived mineral extract, in conjunction with calcium, accounted for the positive effects on bones (Aslam et al., 2013). *Lithothamnion* spp. belong to the order Corallinales. Seaweeds belonging to this order are characterized by the presence of calcium carbonate in their cell walls.

Osteoporosis is a condition characterized by bone deterioration and low bone mass, which increases the risk of fracture. The condition is associated with an imbalance between the formation and resorption of bone tissue. Osteoblasts are cells responsible for bone synthesis, while osteoclasts are cells that break down bone tissue. Several compounds from seaweeds have been shown to induce osteoblast formation, inhibit osteoclast formation or both.

Fucoidan, a polysaccharide found in brown seaweeds, significantly induced osteogenic differentiation—which gives rise to the formation of osteoblasts—in vitro (Cho et al., 2009; Park et al., 2012). Fucoxanthin, a carotenoid found in brown seaweeds, inhibited osteoclast differentiation—a process that gives rise to osteoclasts—in an in vitro study by Das et al. (2010).

Kelp and coralline seaweeds growing on a rock in the intertidal zone.

In an in vitro study, a marine algae–derived calcium supplement called AlgaeCal increased mineralization and proliferation of osteoblast cells (Adluri et al., 2010). In addition, a multi-mineral supplement called Aquamin, sourced from the red seaweed *Lithothamnion corallioides*, increased mineralization in osteoblast cells in vitro (O'Gorman et al., 2011).

In a clinical trial on fifty-eight postmenopausal women with a mean age of eighty, one group was given 900 milligrams per day of calcium derived from heated oyster shell and seaweed, one group was given 900 milligrams of calcium carbonate and one group was given a placebo. Lumbar spine and radial bone mineral density was measured, and researchers found that these indicators increased significantly, compared with the placebo group, in the group taking the calcium sourced from oyster shells and seaweed, but did not significantly increase in the group taking the calcium carbonate (Fujita et al., 1996). In a follow-up clinical trial in 2000 by Fujita et al., an additional subject group was added that received oyster shell–derived calcium without the seaweed component. Only the group receiving calcium derived from both oyster shells and seaweed showed a significant increase in bone density compared with the group receiving the placebo. These results suggest that seaweed-sourced calcium has a superior availability.

SEAWEED AND ENDOCRINE, SEXUAL AND REPRODUCTIVE HEALTH

Seaweeds have been used traditionally in coastal cultures as a male sexual tonic and to promote fertility and reproductive health in women. The brown seaweed *Undaria pinnatifida*, called wakame in Japan and *miyeok* in Korea, is used as a medicinal food. In Japan, it is taken to support women's health: to help regulate the menstrual cycle and purify the blood after childbirth. In Korea, it is made into a soup that is eaten daily for four weeks after childbirth to improve the mother's milk and restore vitality (Ii, 1999). In parts of the Caribbean, the red seaweed *Gracilaria* spp. is used in a male virility drink to increase sexual desire and enhance sexual stamina and performance.

Iodine-rich seaweeds nourish the thyroid gland, which plays a central role in regulating sex hormones. Also rich in phytoestrogens, seaweeds can help ease symptoms of premenstrual syndrome and menopause. The abundance of vitamins including folate (the natural form of folic acid) and bioavailable minerals including calcium, magnesium, iron and rare trace minerals makes seaweed an excellent support for women during pregnancy, breastfeeding, menstruation and menopause. Sufficient iodine intake during pregnancy is critical to the healthy development of the fetal brain. Additionally, sodium alginate from brown seaweeds safely removes toxins associated with birth defects and cancer.

In a study on eighty-four male patients with mild to moderate erectile dysfunction, testosterone at the lower limit of normal and reduced libido,

Winged kelp (*Alaria marginata*) hanging to dry in my seaweed shop. Winged kelp is sometimes called "wild wakame" because it is closely related to Japanese wakame (*Undaria pinnatifida*).

subjects were given a supplement made with the brown seaweed *Ecklonia cava*, the plant *Tribulus terrestris* and chitosan (a compound made from the exoskeleton of crustaceans). Results showed significantly increased testosterone levels and improved libido (Russo et al., 2016).

Consuming fucoidan-rich brown seaweeds could also be useful in the prevention of uterine fibroids. In a study by Chen et al. (2018), fibroid tumour cell growth was inhibited by 50 per cent in forty-eight hours after treatment with fucoidan in vitro. In a follow-up in vivo study by the same authors, fucoidan treatment significantly inhibited tumour growth, supporting their in vitro findings.

Some studies have shown shorter menstrual cycles and elevated estrogen levels to be associated with an increased risk of some estrogen-dependent cancers. In a clinical study on healthy postmenopausal women, dietary seaweed was shown to favourably affect estrogen and phytoestrogen metabolism (Teas, Baldeón et al., 2009). In another study, three premenopausal women with menstrual cycle abnormalities and/or a history of menstrual-related diseases were given the brown seaweed *Fucus vesiculosus*. The results showed an increase in the length of the menstrual cycle and anti-estrogenic effects (Skibola, 2004).

SEAWEED AND METABOLIC HEALTH

The thyroid gland is responsible for regulating metabolism. Sufficient dietary iodine is essential to healthy thyroid function. Low iodine leads to hypothyroidism and goitre (enlarged thyroid); hypothyroidism is associated with a sluggish metabolism, weight gain, fatigue and mood disorders. Seaweeds are a rich source of iodine, which they store in organic molecules, greatly enhancing bioavailability.

Many studies have explored the role of seaweed fibres in managing obesity, one component of metabolic syndrome. Metabolic syndrome is characterized by a set of symptoms that dramatically increase the risk of developing CVD, type 2 diabetes and stroke, including excess abdominal body fat, hyperglycemia, hypertension and low levels of HDLs. The alarming rise in obesity globally is currently considered a public health epidemic.

A diet high in fibre is associated with a decreased risk of obesity. In particular, soluble fibre has specific anti-obesity activity. Fibre is the largest component of seaweed's biomass, including an abundance of soluble fibre (Dawczynski et al., 2007). In addition, magnesium, an abundant mineral in seaweeds, is essential for healthy metabolism and is suggested as a potential weight loss treatment. Moreover, seaweeds are an abundant source of micronutrients, and associations between micronutrient deficiency and obesity have been identified in various populations (Astrup and Bügel, 2010). The anti-inflammatory activity of seaweeds can also be helpful for managing obesity, which is associated with chronic mild inflammation (Trayhurn and Wood, 2005).

A number of studies show the anti-obesity effects of sodium alginate, a dietary fibre from brown seaweeds that is the active ingredient in commercially available weight loss supplements. Sodium alginate has been shown to modulate appetite and energy uptake by reacting with gastric acid to form a gel that slows gastric emptying, stimulates gastric stretch receptors and reduces nutrient absorption in the intestine, which is beneficial for weight loss (Dettmar et al., 2011). In a twelve-week clinical trial on ninety-six obese subjects, one group was given an alginate supplement, administered as a beverage, three times per day before meals, and one group was given a placebo. The group taking the alginate beverage had greater weight loss than the placebo group due primarily to a reduction in body fat (Jensen, Kristensen and Astrup, 2012). In a clinical trial by Pelkman et al. (2007), an alginate-pectin beverage was given twice daily, after meals, to nondieting overweight and obese women. Compared with the placebo, women taking the alginate-pectin beverage after meals had a significant reduction in food intake. In another study on human subjects, sixty-eight men and women were given sodium alginate before meals for seven days. Results showed a 7 per cent reduction in mean daily energy intake, including significant reductions in carbohydrates, sugar, fats and protein. The authors stated that the mechanisms for the observed moderation of appetite appeared to be delayed gastric emptying and an inhibition of nutrient uptake in the small intestine (Paxman et al., 2008b). The same authors report that the uptake of glucose, triacylglycerols and cholesterol increase with increasing body fat. Their results in a crossover study with human subjects suggest that taking a 1.5-gram dose of sodium alginate may restore uptake of glucose and cholesterol to the levels of healthy subjects (Paxman et al., 2008a).

Fucoidan is another fibre found in brown seaweeds that is used in commercial weight loss supplements. In a study on obese diabetic mice, the administration of fucoidan gradually reduced body weight (Jeong et al., 2013). Fucoxanthin, also from brown seaweeds and used in weight loss supplements, has been shown to inhibit weight gain of white adipose tissue (Maeda et al., 2007). In a clinical trial on premenopausal obese women, an oral administration of fucoxanthin and pomegranate seed oil resulted in significant reduction of body weight, waist circumference and liver fat compared with the placebo (Abidov et al., 2010). Phlorotannins, found in all three groups of seaweeds but in the highest amounts in brown seaweeds, also exhibit anti-obesity effects, including the inhibition of lipid accumulation (Chu and Phang, 2016).

Studies on whole seaweeds have also shown anti-obesity effects. In a study on rats fed a high-fat and high-sugar diet, the administration of 5 per cent of total dietary intake of the green seaweed *Ulva ohnoi* resulted in a reduced total body fat mass of 24 per cent after eight weeks (Kumar et al., 2015). Additionally, in a two-month-long clinical trial by Teas et al. (2009), female subjects had reduced waist circumference after consuming 4 grams per day for the first month and 6 grams per day for the second month of the brown seaweed *Undaria pinnatifida* (wakame).

SEAWEED AND MENTAL HEALTH

The calming effect of seaweeds has been employed for centuries, and their capacity to alleviate stress, restore vitality and improve mood is widely recognized. Seaweeds, with their abundance of nutrients, nourish the nerves and muscles and can help combat fatigue and energy depletion. Moreover, seaweeds are rich in bioavailable iodine, which promotes healthy thyroid function, and a sluggish thyroid is associated with fatigue and depression. Melatonin, a hormone involved in the sleep-wake cycle and widely reported to help induce a good night's sleep, is present in significant amounts in seaweeds.

In an in vivo animal model study, extracts from the phlorotannin-rich brown seaweed

The Power of Bathing with Seaweeds

When I met Christine Hopkins at a wellness centre we both worked at in Sooke, BC, we quickly realized we had a mutual love for seaweed, and we less quickly realized that we were rural next-door neighbours. Having identified each other as fellow "seaweed heads," we began travelling to seaweed educational events together to quench our thirst for knowledge.

It was Christine who removed the doubt in my mind that bathing in seaweeds can be just as powerful as eating them. A few years before we met, Christine had learned from a friend that bathing in seaweeds was very therapeutic, reducing stress, improving circulation, easing pain, eliminating toxins, providing better sleep and nourishing the skin. Having close access to our local beaches, she began bathing in seaweeds five to six times per week. At the time, she was taking the drug Synthroid, a synthetic form of thyroid hormone, to treat hypothyroidism (an underactive thyroid). She was told by her doctor that she would need to take the drug daily for life. After she began regularly bathing with seaweeds, she began to experience symptoms of *hyper*thyroidism: a racing heartbeat and being too wired at night to sleep.

She told her doctor about her new symptoms and she asked if Christine was doing anything different, particularly if she had changed her diet in any way. She answered, "No, not that I can think of" at first, but then something prompted her to mention the seaweed baths. Her doctor wondered if they could be responsible for the symptoms because of the rich iodine content in seaweeds and the frequency of the baths.

Her doctor, who is more open-minded than some, agreed to closely monitor Christine's thyroid levels while she continued with her seaweed baths, and to slowly decrease the dosage of Synthroid. Within two and a half months, Christine was able to completely discontinue the synthetic thyroid medication. At that time, Christine was not eating seaweed. That was over fifteen years ago now, and she has remained off Synthroid.

Christine Hopkins with feather boa, teaching at a seaweed workshop.

Nowadays, Christine eats lots of seaweed, in addition to continuing her seaweed baths and using seaweed in her beautiful gardens. She went on to develop her own thalassotherapy (bathing in seaweeds) treatments, and we co-teach several local seaweed workshops each season. Upon healing herself with seaweeds, her passion for them was ignited.

Chondracanthus exasperatus, Turkish towel, is a natural exfoliant.

Ecklonia cava showed significant anticonvulsive and sleep-inducing effects, and the authors suggest its potential use in the treatment of anxiety and insomnia (Cho et al., 2012).

In a study conducted between 2008 and 2011, five hundred employed Japanese adults aged twenty to seventy-four were assessed for diet and depressive symptoms. Results showed a clear relationship between seaweed consumption and a lower incidence of depressive symptoms (Guo et al., 2019). Furthermore, an epidemiological study on 1,745 pregnant women in Japan showed seaweed consumption to be independently associated with a lower prevalence of depressive symptoms (Miyake et al., 2014).

SEAWEED AND THE SKIN

Skin, the largest organ in the human body, is our first line of protection from outside harm. Seaweeds have been used to cleanse, tone, nourish and heal skin for millennia. We can absorb many nutrients from seaweeds transdermally, through our skin. When we bathe with seaweeds, their intracellular fluid is released upon exposure to fresh water; the difference in osmotic pressure between salt water and fresh water causes the cell membranes to burst. Nutrients are then available for the skin to absorb.

Today, seaweeds are used to hydrate and nourish the skin, promote collagen and elastin production, improve tone, clear blemishes and soothe sunburns and dry skin. Some seaweeds, such as Turkish towel (*Chondracanthus exasperatus*) and Turkish washcloth (*Mastocarpus* spp.) are natural exfoliants, gently removing dead cells. Seaweeds reduce inflammation and remove lactic acid, making them ideal to bathe in after rigorous physical activity. Seaweeds are also an excellent source of tocopherols (vitamin E), an antioxidant and important source of protection for the skin and eyes against damage caused by UV exposure. Tocopherols also inhibit lipid peroxidation, the oxidative degradation of lipids that can cause tissue damage.

Applying seaweeds to areas affected by chronic skin conditions such as psoriasis, eczema and rosacea is soothing and healing. The red seaweed *Mazzaella splendens* (rainbow seaweed) is especially recommended as a topical treatment for chronic skin conditions.

IN SUMMARY

Seaweeds are an unparalleled source of nourishment and provide comprehensive support for the body's major systems.

The Voice of Yearning

I love seagulls. People always seem surprised when I proclaim that they are my favourite bird, but it's true. And why not?

Seagulls are beautiful birds. Their lower body and tail feathers are a pure, snowy white, seamlessly transitioning to the smooth, glossy, dolphin-coloured grey of their sides, highlighted with crow black. And the speckled ones! As if an artist came along and sprayed their plumage with a flick of their brush.

I love how seagulls always choose the most artistic place to land. It is as if they are working as a model for a professional photographer.

KATE WOODS PHOTO

I love how seagulls always choose the most artistic place to land. It is as if they are working as a model for a professional photographer. You will often see them on top of a dock piling, a sailboat mast or the outer railing of a boat's deck. I once saw a seagull perched atop a sign at the beach that read "All animals must be on leash." Perfect.

Despite these charming qualities, there are two reasons in particular why seagulls are my favourite bird. First, because their mere presence signifies a beach nearby. Second, the call they make is ethereal. Its cadence is tragic: sad and beautiful. It is rivalled, in my ornithological experience, only by the loon. I am not speaking of the squawk seagulls are famous for making when competing for an abandoned french fry. I am speaking of that call they make while flying alone that can pierce through a lonely morning fog, straight into the heart.

This call, *the seagulls' call*, is the voice of yearning. It is the sound of a lover searching for their soulmate, a mother searching for her child, a mermaid searching for the sea. It is an eerie, impassioned call that strikes through to the part of *me* that yearns.

Phlorotannins

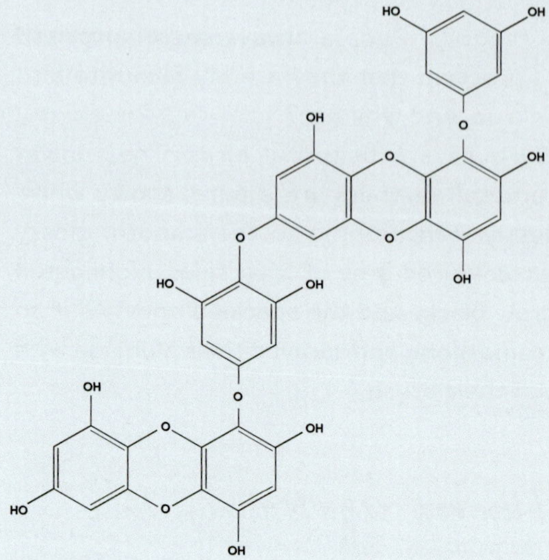

Structure of a dieckol molecule, a phlorotannin present in the brown seaweed *Ecklonia cava*. This seaweed, which is a kelp, is the richest known source of phlorotannins. Dieckol is one of the most powerful antioxidants and has many biological functions beneficial to human health.

Phlorotannins are a group of phenolic compounds (compounds containing hydroxylated aromatic rings) that are unique to algae. They provide structure for cell walls and protection from UV radiation and serve other biological functions. In macroalgae, phlorotannins are found in the highest amounts in the brown seaweeds, but they are also present in some red and green seaweeds. Phlorotannins are also present in diatoms, a group of brown microalgae.

Phlorotannins are one of the most studied groups of marine biologically active compounds, and their antioxidant activity, which accounts for much of their potential in promoting human health, is among the most powerful known. Some studies on phlorotannins have shown an inhibitory effect on angiogenesis in human cancer cells, an essential component of metastasis that is the primary mechanism for cancer spreading from one area to another. Additionally, phlorotannins have shown antidiabetic, antiallergic, antibacterial (including against MRSA-resistant bacteria), antiviral, antifungal, anti-inflammatory, neuroprotective and radioprotective activity. No toxicity has been observed from phlorotannins (Hwang, Terada and Shin, 2008).

A Recurring Dream

When I was in my early twenties, I volunteered with a non-profit group to help orcas. Specifically, this group was collecting data and trying to get official protection for a very small and unique orca habitat: a place called Robson Bight.

Robson Bight is a very small area of water, but what is underneath the water is what attracts the orcas. The bottom of Robson Bight is made up of smooth rocks all similar in size and rises steeply from deep water up to dry land, where it quickly segues into western red cedars and temperate rainforest.

For reasons unconfirmed by science, orcas gather there, rubbing their bellies along the smooth rocks in the shallows. Personally, I think it *feels* good. I fully enjoy a hot rock massage, so why not orcas?

My volunteer position involved camping on a remote island called Cracroft in Johnstone Strait, a body of water located between northern Vancouver Island and the long inlets of the mainland. It is a body of water best known for how fast it can become treacherous and as a permanent home of the Northern Resident Killer Whales (NRKW). When the tide is below a certain point, Cracroft is a single island. When the tide rises above that point, the island is divided into West Cracroft and East Cracroft. The island, at least at the time I was there in the summer of 1999, had no permanent human inhabitants, no infrastructure, and lots and lots of bears. Our camp for the summer volunteer season, which is when orcas belonging to the NRKW population gather to rub their bellies on the smooth beach rocks, was on West Cracroft Island.

On alternating days, I would hike, alone, for forty-five minutes along a ridge that lined the top of the island. I wasn't allowed to have my dog, Willet, with me, as the safety and comfort of wildlife was the number one priority of this group. I had spent plenty of time in the bush among bears while tree-planting, but never without the protective safety of my dog. For perhaps the first time, I found myself thinking often of bears while hiking that ridge with the kind of vigilance that adrenalin helps to maintain.

The hike culminated in a rocky bluff with a perfect view of Robson Bight. On that perch, I would sit with my clipboard for the day, recording how many orcas were in the area we were lobbying to get protected as a reserve, and how many fishing and recreational boats. Just before sunset, I would start my hike back to camp.

On the other days, I got to drive the eighty-horsepower hard-bottomed zodiac. On my training days, I was taught to go very fast, around eighty kilometres per hour, while watching like a hawk for objects in the water such as floating wood. I quickly learned to avoid floating debris spotted at the last second at breakneck speeds. The goal was to be able to get places quickly and safely.

The role of the boat person was to protect the sanctity of the orcas' special rendezvous from the encroachment of humans. The boat person would receive a radio call from the volunteer on the bluff, who had a bird's-eye view of the entire area, telling them where the boat that was entering the proposed reserve was located. The boat person would then motor there as quickly and safely as possible and kindly ask the driver if they wouldn't mind staying outside the perimeter of the proposed orca reserve.

One night I was inside my tent, which was pitched on the flat rocks leading to the shoreline, just above the high tide mark. It was a dark black moonless night. I woke up from a deep, dreaming sleep to the sound of whales breathing: *Pffffffff... hoooooooo!* I heard the loud, distinct sound of air being forced through a blowhole and then the efficient sucking in of the next breath. These sounds were very loud, very close and the most beautiful music to my ears.

I felt as though I was living out, in real life, a recurring dream I had had spanning decades. In this dream, I awake in a black and moonless night to the distinct sound of dolphins breathing. I hear the whistle of air being

I could feel them in my chest. I walked to where the moon had persuaded the tides to go, as if in a trance. They were so close now, and all around me. The night was so dark that still I couldn't see them. I could only hear them. A moment later, only a few feet offshore, I made out the white mist of one whale's warm breath condensing in the cold night.

I walked to where the moon had persuaded the tides to go, as if in a trance. They were so close now, and all around me.

efficiently expelled then drawn in through their blowholes. I walk longingly and instinctually toward the ocean and the source of the breathing: breathing that seems to permeate my soul. The night is so black that I can't see; I only follow the sound of the breathing.

Back on Cracroft Island, I crawled out of my warm sleeping bag into the cool, moist night air of the Pacific Northwest, air infused with the salt of the sea and the life force of the ancient temperate rainforest. The breathing was very loud now and coming often. I knew there was a pod: a family of whales. The sounds were so present,

Back on the shore of my dream, I slowly enter the black ocean. Still, I can't see the dolphins, but the sound of their breathing consumes me, and I can feel their presence all around. The black ocean at night, unlike the blue ocean of day, holds the secret I am looking for: the ability to breathe under water. I walk farther into the black ocean, pulled by an ancient force . . . by an irresistible and passionate urge to go where I truly belong. I feel the tranquil sensation of coming home. As I prepare to dive into the black ocean, I see my future flash before my eyes: an underwater home, among dolphins.

ALGAE

A GLOBAL PERSPECTIVE

"People ask: Why should I care about the ocean? Because the ocean is the cornerstone of earth's life support system, it shapes climate and weather. It holds most of life on earth. 97 per cent of earth's water is there. It's the blue heart of the planet—we should take care of our heart. It's what makes life possible for us. We still have a really good chance to make things better than they are. They won't get better unless we take the action and inspire others to do the same thing. No one is without power. Everybody has the capacity to do something."

—SYLVIA A. EARLE

Our Planet, Our Collective Responsibility

The earth is facing an environmental crisis. Climate change, loss of biodiversity, ecosystem destruction and pollution all point toward an inhabitable planet if we continue on the current trajectory. Despite the world's top scientists telling us that the human race will not survive without radical change, we continue on for the most part with "business as usual."

I believe that if there is to be a planet healthy enough for my two daughters and future grandchildren to enjoy the way I have, individuals, businesses and policymakers will all have to work together to be part of the solution, and the lifestyle that currently exists in the developed world will have to be completely reimagined. Convenience and excess will have to be exchanged for values such as simplicity, quality time spent together and an intimate connection with our food and water, and, by extension, our environment. Solutions to the global environmental crisis come down to a fundamental shift in values because a system where the health of humans and our planet is at odds with economic health is fundamentally flawed.

Environmental lawyer and climate activist Gus Speth puts it this way:

> I used to think the top environmental problems were biodiversity loss, ecosystem collapse and climate change. I thought that with thirty years of good science we could address those problems. But I was wrong. The top environmental problems are selfishness, greed and apathy . . . and to deal with those we need a spiritual and cultural transformation. (WineWaterWatch.org, 2015)

Feeling overwhelmed by the current state of our environment can cause extreme feelings of helplessness, hopelessness and chronic anxiety, particularly for our youth. The story of "The Star Thrower" reminds me that our individual actions *do* matter. I find comfort in it. The original story, written by Loren Eiseley, has inspired many variations since it was published in 1969. This is the story as I like to tell it, in my own words:

> A boy walked along a beach where hundreds of starfish had been stranded by the ebbing tide. An old man sat on a log and watched as the boy picked up a starfish and threw it back into the sea. The boy continued picking up starfish and throwing them back into the sea as the man watched, but it didn't seem to make any noticeable difference to how many were still stranded on the beach. The man approached the boy just as he was about to throw another starfish back into the sea. The man asked the boy why he was doing that when it was apparent that he would never be able to make a difference to the hundreds that still lay there stranded. The boy answered the man, "It makes a difference to this one."

As one whose lifestyle and livelihood depend upon the health of the oceans, I have a deep responsibility for stewarding the ocean to the absolute best of my ability, both locally and globally. Climate change, pollution, ecosystem destruction, overfishing and industrial fish farms are at the top of the list of ocean issues in dire need of solutions. Healthy microalgae (phytoplankton) populations, thriving coastal seaweed ecosystems and robust kelp forests are crucial to the health of the oceans. What's more, algae are a source of potential solutions to some of the ocean's—and the world's—most critical environmental issues.

Our most fundamental relationship is with the planet that sustains us and gives us life. Individual actions can have a ripple effect and spread like rings from a starfish thrown with love into the sea.

A Fish out of Water

There is an inherent and deeply isolating loneliness that exists inside when you have a steady and unrelenting yearning telling you that you belong in a different element than the one you were born into: spending one's whole life as a fish out of water.

Surfing and diving among the kelp forests of the Pacific Northwest for the better part of my adult life have provided some relief from this feeling.

Wherever I have travelled, I have dipped myself into the salty ocean. Slipping sneakily into harbours on America's Pacific coast, entering the warm, black ocean of Honduras, Costa Rica and Nicaragua at night, cruising with lionfish in the Indian Ocean off Kenya, and cleansing my spirit in the icy Atlantic of Nova Scotia in winter.

The salt is always a comforting confirmation of the place where my spirit resides.

The salt is always a comforting confirmation of the place where my spirit resides.

CHRIS ADAIR PHOTO

Intertidal seaweeds with a tide pool.

The Central Role of Algae in Global Ecological Health

"With every drop of water you drink, every breath you take, you're connected to the sea. No matter where on earth you live."

—SYLVIA EARLE, TED CONFERENCE

The world needs algae. Algae produce an estimated 50 to 80 per cent of global oxygen and absorb an estimated one-third of atmospheric carbon dioxide. Algae are the primary producers of the ocean ecosystem, which covers two-thirds of our blue planet. They convert sunlight to energy, which travels from tiny plankton to the mighty orcas at the apex of the oceanic food chain.

Algae are also major contributors to cloud formation. When algae cells are damaged or die, they release dimethylsulphoniopropionate, which breaks down into dimethyl sulphide (DMS), the most abundant biological sulphur compound in the atmosphere. DMS reacts with oxygen in the air to produce aerosol sulphates, which act as nuclei for cloud condensation. This process is critical to the regulation of climate.

The oceans support an enormous amount of biodiversity. According to the United Nations Educational, Scientific and Cultural Organization (UNESCO), an estimated 50 to 80 per cent of life on earth exists in the oceans. The diversity of this life is astounding, extending from microscopic algae and bacteria all the way up to the magnificent blue whale, the largest animal to ever live on earth. This plethora of life is made possible by algae.

Diversity in the oceans is exemplified not only by the sheer number of species, but also by the different types of ecosystems that support high biodiversity. For example, coral reefs and kelp forests, dependent upon microalgae and macroalgae respectively for primary production, have the highest biodiversity in the oceans and are among the most biodiverse ecosystems in the world. Coral reefs, incredibly, exist in nutrient-poor tropical waters, yet support thousands of different species, including an estimated 25 per cent of all marine fish species. Corals are tiny animals that live in large colonies. They live in a symbiotic relationship with zooxanthellae, which belong to a group of microalgae called dinoflagellates. The corals are dependent upon these algae for nutrients and cannot survive without them. Coral reefs, created over thousands of years, are currently dying at an alarming rate due to a process called bleaching. Bleaching occurs when algae leave coral as a stress response to increased water temperature, breaking their symbiotic relationship. Coral can only survive a short time without algae, and eventually die of starvation. According to NOAA, the longest, most widespread and destructive bleaching event occurred from 2014 to 2017, when 75 per cent of the world's coral reef systems experienced bleaching-inducing temperatures and 30 per cent experienced temperatures that caused bleaching-induced mortality.

Kelp forests, one of the most productive ecosystems in the world, exist in nutrient-rich temperate coastal waters. Macroalgae, specifically large kelp belonging to the brown seaweeds, provide structure, biomass and food for herbivores. The bottom of a kelp forest is covered with diverse seaweed species, including kelp and other brown seaweeds and green and red seaweeds, including coralline algae. A dynamic ecosystem, kelp forests can appear and disappear based on weather, seasonality, predation and human activity. Like coral reefs, kelp forests are in a global state of decline with warming ocean temperatures cited as the primary cause.

In addition to algae's role as primary producers, they also purify the ocean of toxins. Seaweeds absorb compounds such as heavy metals from the surrounding waters and convert them into less harmful molecules that are essentially biologically inert. Seaweeds are used specifically as "environmental scrubbers" to decontaminate severely polluted waterways.

Algae play a central role in the ecology of the earth. From the air we breathe to the regulation of climate to our primary source of energy, algae, earth's oldest organisms, are essential to the existence of life.

Kelp forest ecosystems are teeming with life.

CHRIS ADAIR PHOTO

Slow Fish . . . Finding My Crew

> ***"I arise in the morning torn between a desire to improve (or save) the world and a desire to enjoy (or savor) the world. This makes it hard to plan the day."***
>
> —E.B. WHITE, *NEW YORK TIMES* INTERVIEW

One spring day in 2015, while peddling my local, hand-harvested seaweed at Wild Mountain, a new local restaurant in Sooke, I met Brooke Fader and was introduced to Slow Food. Brooke is the convivium leader of Slow Food Vancouver Island and Gulf Islands and a founding member of Slow Fish Canada.

In the fall of 2015 Brooke sent me an application to become a delegate at Terra Madre Salone del Gusto, the biannual international Slow Food gathering that happens in Turin, Italy. At her insistence that I would be "perfect" for the position, I sent the application in half-heartedly, knowing next to nothing about Slow Food and precisely nothing about Terra Madre. I was accepted and asked to speak at the Slow Fish area of the gathering in September 2016.

Attending a Slow Fish gathering is like lighting your soul on fire: the hopelessness and despair at the magnitude of the challenges we face transform into hope and belief.

Kris Barnholden (left), Ingrid Jarrett (right) and me after presenting a Taste Workshop at Terra Madre Salone del Gusto in Turin, Italy, in 2018. The magic of Slow Fish is the network: the connections made and the knowledge shared.

Terra Madre dramatically transformed any preconceived notions I had about how to heal the global food system, the crisis in the oceans and just how many people are working toward this goal. The forward momentum at that gathering of thousands of people, including seven thousand delegates from 143 countries, working in a diverse way toward a common goal, ignited something inside me that I felt compelled to explore further. When I heard that Slow Fish was gathering in San Francisco in 2018, I juggled being a single mom, finances and work and got myself there.

Slow Fish is an international grassroots network of people passionate about and committed to a healthy ocean ecosystem . . . one that includes people. Slow Fish does not believe in a one-size-fits-all solution but rather in local solutions that are as diverse as the ocean itself. We cannot survive without the sea.

The connections I made at Slow Fish San Francisco and Slow Fish Terra Madre have been

The 2018 Slow Fish gathering was kicked off with a variety of oysters offered by small-scale oyster farmers at the TwoXSea warehouse at Fisherman's Wharf in San Francisco.

life-altering. These were true heroes promoting Slow Fish values, which speak to the core of what I believe in:

- Good: delicious, fresh, local, prepared well, honoured
- Clean: a healthy ocean and healthy ecosystems, unadulterated food from the ocean, free of chemical additives and preservatives
- Fair: a livable wage that preserves the dignity, lifestyle and cultures of small-scale fishers and harvesters

Slow Fish assumes that the will to act stems from fierce and intrinsic joy; joy as primal as sucking the briny juice of an oyster from the smooth pearly white of its inner shell. Attending a Slow Fish gathering is like lighting your soul on fire: the hopelessness and despair at the magnitude of the challenges we face transform into hope and belief. Most importantly, it fills up the fuel tanks necessary for each of us to go back to our corners of the world and keep on fighting fiercely for *good, clean and fair*.

Algae and the Challenge of Climate Change

"I will believe people really care and are serious about climate change when they start talking and caring about phytoplankton [microalgae]."

—MICHÈLE MESMAIN, FORMER COORDINATOR OF THE INTERNATIONAL SLOW FISH MOVEMENT

ALGAE, CO_2 ABSORPTION AND OCEAN ACIDIFICATION

The oceans play a critical role in regulating climate and keeping global temperatures down. They absorb most of the excess heat (an estimated 90 per cent or more) caused by greenhouse gases. Marine algae also absorb an estimated one-third of atmospheric carbon dioxide, the most prevalent greenhouse gas aside from water vapour.

Unfortunately, the absorption of the excess carbon dioxide currently in our atmosphere is causing changes in the chemistry of the oceans, including increasing acidity and an associated decrease in carbonate ions, which many marine species depend on for survival. This process is called ocean acidification. While scientists are just starting to uncover the effects of ocean acidification, we know that it is having a negative impact on many marine organisms and that calcareous organisms (those with a large component of calcium carbonate) are especially sensitive. Of particular concern is the negative effect on coccolithophores, a very abundant group of calcareous microalgae that play a major role in the biogeochemical cycles that regulate global climate. Other organisms known to be negatively affected by ocean acidification include echinoderms (e.g. starfish), cnidarians (e.g. coral), crustaceans, molluscs and fish.

Seaweeds (macroalgae) have an alkalizing effect on their surrounding waters, just as they do in our bodies, and can positively influence the pH of the water where they grow. Oysters are particularly sensitive to small increases in acidity, and for this reason some oyster farmers grow seaweeds in order to lower acidity and improve water quality in their oyster farms.

REDUCING METHANE EMISSIONS FROM LIVESTOCK

Raising livestock on an industrial scale is among the top contributors to global warming. Its climate impacts include deforestation to create pastures for livestock, methane production (which has an estimated warming effect twenty-five times higher than carbon dioxide) released by ruminants during digestion and from their manure, and nitrous oxide production (which has an estimated warming effect 265 times higher than carbon dioxide) from the use of fertilizers on feed crops.

In the early 2000s, Joe Dorgan, a Canadian farmer from Prince Edward Island, noticed that his cows who had access to the seashore appeared healthier and produced more milk than his inland herd. He wondered if the nutrient-rich seaweed they grazed on was the cause, and so he started collecting washed-up seaweed and feeding it to his inland herds. Dorgan began to notice the same effect on his inland cows: they seemed healthier and produced more milk. He believed that using seaweed as an additive in livestock feed, as it has been used traditionally for thousands of years, could have potential as a business and he wanted to quantify the benefits of seaweed to grazing livestock. Rob Kinley, at the time a research scientist at Dalhousie University, studied Dorgan's seaweed-consuming cows and discovered that not only was seaweed increasing the efficiency of their digestion, but it was also *decreasing* methane production by 12 per cent. The energy saved went toward milk production, which increased. Kinley wondered if different species of seaweed might have an even greater effect on methane production. With this in mind, he teamed up with phycologists and ruminant nutrition experts in Australia, where an estimated 10 per cent of all greenhouse gas emissions is methane produced by livestock. The research team discovered that the red seaweed *Asparagopsis taxiformis* had dramatic effects on methane output. In vitro studies showed a 95 per cent reduction in methane output with 5 per cent seaweed administered (Roque et al., 2019). In a study on live sheep, methane output was reduced by up to 80 per cent with 3 per cent seaweed (*Asparagopsis taxiformis*) added to the diet (Li et al., 2016).

An estimated 39 per cent of the greenhouse gas emissions created by raising livestock is methane, which accounts for approximately one-quarter of the world's methane pollution. In turn, livestock-related emissions make up approximately 14.5 to 18 per cent of global greenhouse gas emissions.

Healthy coastal seaweed ecosystems help to reduce the amount of the greenhouse gas carbon dioxide in the atmosphere.

NEW, MORE SUSTAINABLE BIOFUELS

Biofuels, fuels made from oil-containing plants, have been proposed as an alternative to the burning of fossil fuels, which accounts for an estimated 65 per cent of global greenhouse gas emissions. Currently, corn and soy are the major sources of biofuels. Unfortunately, the arable land, clean water and production that go into growing these crops create significant greenhouse gas emissions.

Microalgae can also be used as a biofuel and has substantial advantages: it is not a food crop of humans and livestock, and it can be grown on a relatively small area of non-arable land. It can also be grown using agricultural wastewater containing chemical fertilizers and other agricultural contaminants, which is then purified by the microalgae, helping to prevent the pollution of rivers, lakes and coastal waters. Additionally, microalgae requires few nutrients, grows much faster than conventional crops and has a higher oil productivity than all other current oil crops. The following fuels have been produced with microalgae: biodiesel, biobutanol, biogasoline, methane, ethanol, green diesel and, particularly encouraging, jet biofuel. The Carbon War Room, an international non-governmental organization and think tank working on market-based solutions to climate change, calls sustainable aviation fuels "the most challenging emissions reduction opportunity" as well as "the greatest potential for achieving carbon-neutral growth in aviation" (Hawken, 2017).

IN SUMMARY

Climate change is the most imperative environmental issue of our time, threatening the very existence of the human race. Algae play a critical role in the regulation of climate and are a key source for innovations aimed at combating climate change.

JULY 2019

The Blessing of Cold

Today is one of those rare days when there is no current, and the blades of the bull kelp hang straight down, following their stipe into the green abyss below. On days like these, it is possible to weave deep into the kelp forest without diving under, an impossibility on most days, when the current pulls the blades to form a thick canopy.

I experience the cold as an agent of healing. As the harm of climate change and a rapidly heating world are always fresh in my mind, I experience this frigid water as a global blessing.

My inner core awakens in the frigid waters of the Pacific Northwest. CHRIS ADAIR PHOTO

As I enter the water to begin my harvest, the jolting cold pierces the part of my face around my mask that is exposed, the only part of me not covered by neoprene. It is shocking, and if I didn't have eighteen years of experience, I would never believe that in a few moments the piercing would fade.

Sure enough, the bite soon fades to a comfortable and familiar numbness. I soften to the cold, trust it and let it in. As the beautiful, large, healthy bull kelps come into view through my mask and the frigid water awakens every part of my body, I experience the cold as an agent of healing. As the harm of climate change and a rapidly heating world are always fresh in my mind, I experience this frigid water as a global blessing. The kelp cannot survive without cold water and, perhaps, neither can I.

Marine Plastic Pollution

"Today we use plastic—a material designed to last forever—for products designed to last minutes."

—MATT PRINDIVILLE, CEO OF UPSTREAM

By 2050 it is estimated that there will be more plastic than fish, by weight, in the ocean. The problem is insidious: plastics have been found in the digestive tracts and tissues of marine creatures from birds to whales, from fish to oysters. This versatile and indestructible material can be almost limitlessly recycled and should have been reserved for such uses as life-saving medical technologies and reusable food packaging in a closed-loop recycling system. Unfortunately, society treats plastic as a disposable item. Everything from pre-made salads to shampoo, moisturizer, hand wipes and countless foods, drinks, personal care items, toys and much more comes in single-use plastics: plastic purposely designed to be used only once. Ready-made food and drinks that will be consumed within minutes come in containers that will last generations in our environment and, as research is now showing, will become incorporated into the food chain. When you dispose of plastic, it doesn't actually go away, as is evidenced on beaches the world over.

THE GREAT PACIFIC GARBAGE PATCH

In the north-central Pacific Ocean, midway between California and Hawaii and nearly two thousand kilometres from any coast, there is an unimaginably massive floating patch of garbage called the Great Pacific Garbage Patch (GPGP). In a 2018 study, Lebreton et al. estimated that the GPGP covers 1.6 million square kilometres and contains at least 79 thousand tonnes of floating plastic pollution. It is the largest area of floating plastic pollution on the planet, and some of the plastic collected from the patch is over fifty years old. Lebreton et al. (2018) report that their results suggest ocean plastic pollution within the GPGP is increasing exponentially. In addition, the authors estimate that microplastics, pieces of plastic less than five millimetres in length, make up only 8 per cent of the plastic pollution by weight in the GPGP but 94 per cent of the estimated 1.8 trillion pieces of floating plastic. Pieces this small are extremely difficult to clean up and are much more harmful to living organisms, highlighting the need for clean-up efforts to happen now, before more marine plastic pollution breaks down into microplastics.

While there are many organizations and individuals cleaning plastic debris from beaches, there has been little effort to remove plastic pollution at sea. One notable exception is the valiant effort by Boyan Slat, a Dutch inventor and entrepreneur who at just eighteen years old founded his company, The Ocean Cleanup. After seeing more plastic than fish on a diving trip in Greece when he was sixteen, Slat became determined to clean up plastic pollution in the ocean. He envisioned a massive floating barrier that could trap and collect floating plastic utilizing the energy of the ocean currents. His vision came close to fruition in 2018 when his invention, named System 001, was launched out of San Francisco en route to the GPGP. Although System 001 was not successful at collecting plastic pollution, Slat and his team were able to apply the lessons learned from its failure to create System 001/B, which was launched out of Vancouver, BC, in 2019. This time, Slat's invention, powered by ocean currents, successfully captured plastic debris. Moreover, this new device has also collected microplastics, a discovery described as a "pleasant surprise" by Slat and his team (The Ocean Cleanup, 2019).

The GPGP is still rapidly increasing in size. It is estimated that approximately half of all plastic produced today goes into single-use plastics and that only 9 per cent is recycled globally. Cleaning up plastic pollution while the production of single-use plastics is at an all-time high feels a lot like using a bowl to scoop water out of an overflowing bathtub while the faucet is still running.

MERMAIDS' TEARS

Plastic only breaks down into smaller pieces; it doesn't biodegrade, meaning it doesn't change molecular structure into something that can be reused by nature. Microplastics, defined as pieces of plastic less than five millimetres long, are especially hazardous and have come to be known as "mermaids' tears." They are the perfect size to be ingested. Hundreds of species of animals have shown evidence of plastic ingestion, including whales, seals, seabirds, turtles, sharks, fish, shrimp, corals, crabs and even filter feeders such mussels, clams and oysters. More than a million sea creatures die each year due to the ingestion of plastic. In many cases there is such a large amount of plastic in the gastrointestinal tract that the animal can no longer ingest and absorb enough nutrients to survive, eventually dying of starvation.

Furthermore, microplastics have an affinity for toxins, specifically persistent organic pollutants (POPs). POPs are a group of toxins harmful to humans and animals that resist environmental degradation and can persist for many years in the environment, bioaccumulating in the food chain. POPs include toxins such as dichlorodiphenyltrichloroethane (DDT), PCBs and dioxins. Microplastics are incredibly efficient at accumulating these toxins, typically having concentrations many times greater than the surrounding water. In a study by Mato et al. in 2001, concentrations of PCBs and dichlorodiphenyldichloroethylene (DDE), a chemical formed from the breakdown of DDT, were found to be one hundred thousand to one million times greater on microplastics than in the surrounding seawater. When these toxin-carrying mermaids' tears are ingested, they not only cause hazards to the digestive tract, but can also be poisonous, as the toxic chemicals are transferred from the microplastic to the organism.

Microplastics have been found in water samples and in the tissues of animals from all over the world's oceans, including such remote places as the Southern Ocean and Mariana Trench, the deepest place in the ocean. Primary microplastics are plastics that are less than five millimetres in size when they enter the environment. Textiles made from synthetic microfibres, such as fleece, acrylic and nylon, slough off many tiny pieces of plastic when washed, which drain into the sewage system, where they are too small for most wastewater systems to filter. In a study by Napper and Thompson (2016), an average of seven hundred thousand microfibres were shed in a six-kilogram wash load of acrylic fibre. Microfibres, considered to be the greatest source of primary microplastic pollution, are estimated to be the source of up to 35 per cent of that pollution and have been found in every region of the world's oceans (Henry, Laitala and Klepp, 2019). Two-thirds of clothing today is produced from synthetic fibres. Recently there has been mounting evidence that wear and tear from car tires, which release tiny pieces of plastic into the environment, may be a comparable source of microplastic pollution, with 2017 estimates putting it at 5 to 10 per cent of global marine plastic pollution (Kole et al., 2017).

Nurdles, small plastic pellets that are used in the production of most plastic products, are yet another significant contributor to marine microplastic pollution. Nurdles are typically the size of a lentil and enter the environment through spillage, both during production and in transport. Smaller still are microbeads: tiny beads of plastic used as abrasives in products such as cosmetics, shampoo, toothpaste and cleaners. Like microfibres, these plastic particles are too small to be filtered by sewage treatment plants and end up in rivers, lakes and oceans. Fortunately, the Canadian government banned the production, import and sale of products containing microbeads in 2019, following the lead of the US and the UK. Currently

Plastic pollution or barnacle habitat or both? Now is the time to come together and solve the plastic pollution crisis in our oceans.

the United Nations is urging a worldwide ban on microbeads.

The majority of microplastics, however, are secondary microplastics: small plastic pieces created by the breaking down of larger plastic pollution. It is much harder to study the impact of secondary microplastics on the health of marine organisms and humans because of their variability in composition, size and shape. In all forms, microplastics pose a threat to diverse living organisms. The World Health Organization describes microplastics as "ubiquitous in the environment" and states that they "have been detected in marine water, wastewater, fresh water, food, air and drinking-water, both bottled and tap water" (World Health Organization, 2019).

DISPOSABLE PLASTIC WATER BOTTLES

When I grew up in the 1980s, the notion of bottled water—of *paying* for water in North America, where potable water is widely free and available—seemed ludicrous. However, Coke, Pepsi, Nestlé and other companies predicted and helped drive the exponential growth of the bottled water market, which was worth $185 billion in 2015 and is continuing to rise (Research and Markets, 2018). Worldwide, an estimated twenty thousand plastic water bottles are purchased every second, and the trillions of bottles being disposed of are a major contributor to plastic pollution (Laville and Taylor, 2017). Moreover, the water inside these bottles appears to be contaminated by the very pollution their packaging creates. In a 2018 study by Mason, Welch and Neratko, 259 bottles of water by eleven global brands were purchased from nineteen locations in nine different countries and tested to determine the presence of microplastics. Of the bottles tested, 93 per cent contained microplastics. Ironically but not surprisingly, Mason et al. report that the data suggests the microplastic contamination at least partially comes from plastic water bottle-sourced pollution.

Microplastic pollution is also prevalent in tap water. In a 2018 study by Kosuth et al., 89 per cent of 159 samples of globally sourced tap water had microplastic contamination. However, currently available data suggests that overall microplastic contamination is significantly higher in bottled water, although the level of contamination is highly variable among samples, even samples from the same brand of bottled water or source of tap water. One of the most disheartening facts about disposable water bottles is that they are made out of polyethylene terephthalate (PET), which is one of the most easily recyclable types of plastic, yet the vast majority end up as garbage. Even those that do get recycled usually end up being downcycled and used in products such as carpeting and packaging rather than going back into bottles. Water bottles are one of the most common forms of plastic found washed up on beaches.

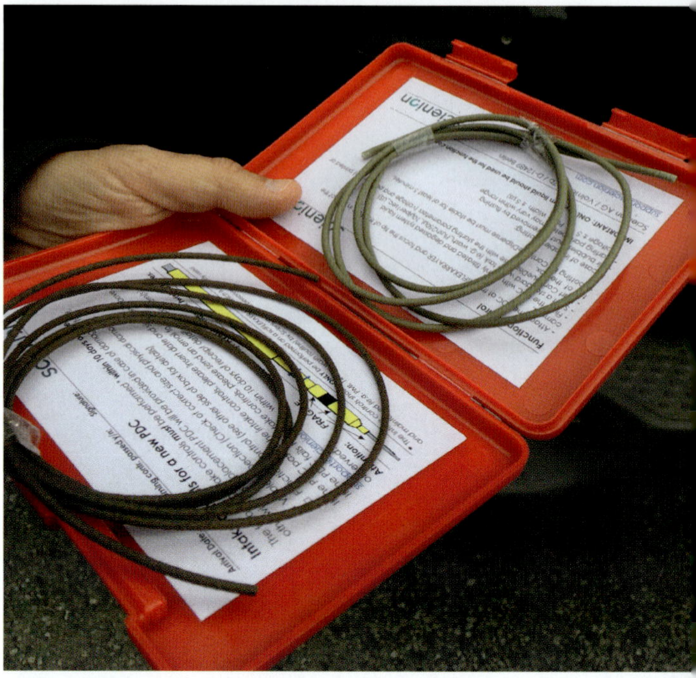

Prototypes of Dr. Hugenholtz's seaweed-based alternative to hard plastic. The top one is made with green seaweed, the bottom with brown seaweed.

SOLUTIONS

The movement promoting an end to single-use plastics is not new, but it has been gaining traction in recent years, with municipal, state and provincial and even federal governments starting to get on board. Many cities have banned plastic bags in retail stores and straws in restaurants, which is a small but significant step. Under Prime Minister Justin Trudeau, Canada's federal government has committed to banning single-use plastics nationwide as early as 2021 and has revealed plans to make manufacturers of plastic products responsible for recycling their plastic waste. Zero-waste and package-free stores are starting to pop up across the country and beyond, offering consumers a shopping option free of single-use plastics.

A few years ago, I had the pleasure of taking Dr. Jeroen Hugenholtz, a professor at Wageningen University in the Netherlands, on a seaweed tour. Hugenholtz was working on developing hard plastic alternatives made with seaweeds. He brought two of his prototypes with him and showed them to me. More than half of their composition was seaweed: one was made with a green seaweed and one with a brown seaweed. Currently, most bioplastics (plastic-like materials made from renewable organic material) are made from derivatives of corn—in particular, corn starch. But unlike corn, algae can be produced without the use of arable land, chemical fertilizers and clean water. Hence, algae as a source for bioplastic can help mitigate the harms of plastic pollution *and* the impacts associated with industrial agricultural monocropping such as greenhouse gas

Husband and wife entrepreneurs and plastic pollution activists Nairn and Paula McPhee opened their business, Zero Waste Emporium, in Victoria in 2018. They offer groceries, cleaning supplies and personal hygiene products completely package free. SALLY MCINTYRE PHOTO

In March of 2019, a young Cuvier's beaked whale was found in the Davao Gulf of the Philippines looking emaciated and vomiting blood. It soon died and forty kilograms of plastic was discovered in its stomach. It was determined that the young whale died of starvation due to the ingestion of plastic. UNESCO estimates one hundred thousand marine mammals die from the ingestion of plastic each year.

emissions, water use, water contamination and soil erosion.

It is hard to find good alternatives to plastic—ones that take into consideration the amount of energy and resources put into production and what the fate of the product will be at the end of its life. For my business, I have switched to plastic-free bags for my retail-sized bags of seaweed. It took me months of researching to find a suitable alternative to plastic. The material I found is made from recycled wood pulp and is food safe and 100 per cent compostable according to both US and European standards. Small changes like these are a necessary step, even though they feel like a drop in the bucket when you realize how embedded single-use plastics have become in our society.

If we commit to action, perhaps in the future the only "debris" washed up on beaches will be natural, such as this tangle of bull kelp.

An additional challenge with bioplastics is creating the technologies and infrastructure for their recycling, which is still in the developmental stages. Despite the challenges, bioplastics offer an important alternative to petroleum-derived plastics. At the heart of the matter, though—and where radical change is critical—is rethinking the concepts of "disposable" and "convenience."

Today you can go to almost any beach in the world and find evidence of the plastic pollution crisis. My own beach, where I was previously fortunate to almost never see plastic, has now succumbed. With the king tides in January 2020, coupled with high swell and wind, plastic bottles, caps and other debris have begun washing up. Today, there is no beach nor any place in the ocean left that is immune to plastic pollution, and fewer unaffected marine organisms each day. I hope that we can come together and make the choice to solve this crisis: individuals, entrepreneurs, innovators, corporations and policymakers. It will take every one of us.

Serenading the Seaweed

When I sing and play to the seaweed, the sound filters through the hanging kelp, bounces off the walls and gently weaves its way back through the long, briny blades.

Inside the seaweed shop, it is always summer. Sometimes it is the thick humidity of a tropical rainforest where the moisture sticks to your skin the moment you enter. Other times, it is the bone-dry heat of the desert that makes your throat tight and thirsty. Often, it is transitioning from one to the other, much like the weather on a Caribbean beach. It all depends on how dry the seaweed is. The aroma, however, is always at low tide.

The seaweed in my shop is of a musical variety. Countless hours have been spent hanging seaweed to a variety of wonderful rhythms. When my daughters were young, the seaweed shop was a place I could head to with my guitar after they were asleep and sing to my heart's content. Essentially a big wooden box with a concrete floor, the shop's acoustics are as lively as a sand flea on a pile of rotting kelp. The natural reverb is akin to being in an old, grand church.

There is no more inspiring place, in my mind, to play music and sing. When I sing and play to the seaweed, the sound filters through the hanging kelp, bounces off the walls and gently weaves its way back through the long, briny blades. Also, the particular seaweed drying at such a time is value-added, because it has been serenaded.

Bull kelp hanging to dry in the seaweed shop.

Eating Sustainably from the Sea

"Love is like seaweed; even if you push it away, you will not prevent it from coming back."

—NIGERIAN PROVERB

One of the most consequential global crises of the twenty-first century is the way that humans produce food: our connection to food and our food systems is, in many places, fundamentally broken. Practices such as industrial agriculture, overconsumption of meat, the inhumane treatment of animals, export crops, bycatch, overfishing and large-scale aquaculture are destructive to the environment, to our societies and to our health. Eating sustainably is a newer concept, and an awareness of how our food is produced is essential to making sustainable food choices. Replacing ecologically and socially destructive practices with systems that work in harmony with nature, and that produce foods that are nutrient dense and health promoting, is necessary for our survival as a species. Seaweeds, packed with life-giving vitamins, minerals and other nutrients and requiring no chemicals or fresh water to grow and harvest, are a healthy part of the solution to repairing our broken food systems.

TRACEABILITY

Traceability is becoming increasingly difficult as more and more of our food is imported from great distances. As a whole, seafood is the most difficult group of foods to trace. According to a 2016 report entitled *Taking Stock* by the non-profit, science-based seafood watchdog SeaChoice, nearly one-third of imported seafood in Canada is untraceable and poorly labelled (Govender et al., 2016). This makes it impossible to determine if it was sustainably produced. In many cases, seafood is caught or farmed in one place, processed somewhere else, often thousands of kilometres away, and then sold to consumers in yet another place. In addition, SeaChoice reports that they were unable to determine where Canada's most unsustainable seafood exports end up due to lack of tracking ability.

BYCATCH

In a 2009 report on global fishing, it was estimated that 40 per cent of the global catch is bycatch (Davies et al., 2009). Bycatch is the incidental capture of non-target species that most often are thrown back into the ocean to die because they are not commercially valuable. Bycatch includes juvenile fish, marine mammals, seabirds and feed species important for commercially valuable species of fish, as well as many other ecologically important species. Problematically, bycatch is unreported, making it difficult to properly manage fisheries. The authors of the report comment that this is a conservative estimate and cite the indiscriminate capture of non-target species in many commercial fisheries to be among the most urgent threats to remaining ocean fish stocks.

OVERFISHING

"We associate the North Atlantic with cod. The motto of Newfoundland used to be 'In cod we trust.' It was a joke, but it was essentially true. But there is no cod anymore. And that's extraordinary. It's all because of either greed or politics."

—SIMON WINCHESTER

The North Atlantic cod fishery collapsed and was closed in 1992. At the time, it was the largest in the world and drove the economy of Newfoundland and Labrador. Although government officials at the Department of Fisheries and Oceans (DFO) announced that the moratorium on cod would last for only a couple of years, the fishery remains mostly closed today, nearly thirty years later. The moratorium resulted in the largest mass layoff in Canadian history, affecting an estimated forty thousand people from Canada's Atlantic provinces, mostly from Newfoundland.

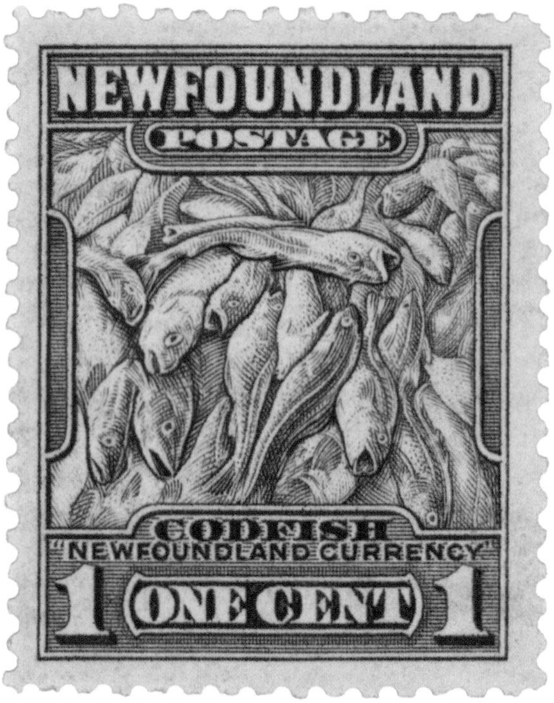

A Newfoundland stamp issued in 1932.

Healthy seaweed ecosystems support healthy fish populations.

In the years just before the fishery collapsed, the small-scale inshore fishers of Newfoundland had said loudly that the cod were gone from their inshore fishing grounds and tried to warn officials at the DFO. Their pleas were ignored as foreign offshore fleets fishing outside of the two-hundred-nautical-mile exclusive economic zone (EEZ), harvesting massive amounts of fish with much more advanced technologies, were landing the biggest catches ever. As the local fishers had warned, the following year there was no cod and no fishery. In a 1989 interview with the CBC, Justin McGrath, a fisherman from Fogo Island, Newfoundland, remembered a time when "the fish would gladly leap into the boat just to get out of the crowded sea." The CBC reported that the disappearance of the cod is considered one of the biggest ecological disasters of the twentieth century.

Dramatic increases in fishing technology, including the use of aircraft, sonar and satellites to locate fish, have allowed for fewer boats to catch significantly more fish. The open ocean, typically anywhere more than two hundred nautical miles from the coast, which is the EEZ of the nearest country, is unregulated. International fishing boats with no connection to the coastal communities adjacent to where they fish do not have the same incentive to fish sustainably as the local communities do and will often fish until the resource is gone and then move elsewhere. Practices like this devastate small-scale coastal fishing communities and are often much worse in waters off the coast of developing nations. In contrast, small-scale fishers who live on the coast are reliant on their local fisheries for their livelihood, lifestyle and food. Often, they succeed many generations of fishing by their predecessors and have both the knowledge and need to fish sustainably and ensure healthy fish populations for future generations.

TROPICAL FARMED SHRIMP

Tropical farmed shrimp (here I use the word "shrimp" to mean both prawns and shrimp) are one of the most unsustainable foods in the world. The developed world's thirst for cheap prawns is driving an industry that is associated with severely destructive ecological practices and extreme social injustice. Shrimp are currently the most popular seafood in the US (Food and Agricultural Organization of the United Nations, n.d.), and 90 per cent of shrimp consumed in the US is imported, with the vast majority coming from Southeast Asia and Central America (Guy, 2017).

Ecologist Boone Kauffman of Oregon State University estimates that shrimp produced on a typical Asian shrimp farm have an estimated carbon footprint ten times greater than beef produced on deforested Amazon rainforest land, primarily because of where they are located (Stokstad, 2012). An estimated 50 to 60 per cent of tropical shrimp farms are located in tidal zones in Asian countries, most of these on land that has been cleared of mangrove forests (Guy, 2017). Mangrove forests are one of the most productive ecosystems in the world, rich in biodiversity, and play a critical ecological role. They act as giant filters to water entering the ocean and as a nursery and habitat to abundant and diverse marine species. They also help protect the land from erosion and damage from hurricanes and tsunamis. Furthermore, they sequester huge amounts of carbon and store it in a way that can contain it for thousands of years if left intact. The destruction of mangrove forests releases significant amounts of stored carbon back into the environment. In addition, shrimp farms on cleared mangrove tidal flats last only an average of five years, at which time contamination by acid sulphate soil and a buildup of sludge render the water uninhabitable to shrimp (Stokstad, 2012). As fish and other marine life that live and breed in mangrove forests disappear with their destruction, coastal fishers and their families that depend on healthy fish, crab, wild shrimp and other marine species for food and livelihoods can suffer food insecurity and displacement. Oftentimes, the privatization of lands by shrimp farms blocks access to traditional fishing grounds.

The tropical shrimp farm industry is also associated with direct human rights abuses. In a 2003 report entitled *Smash & Grab: Conflict, Corruption and Human Rights Abuses in the Shrimp Farming Industry*, the Environmental Justice Foundation documented reports of human rights abuses including murder, sexual abuse and rape, land seizure, child labour and forced labour. According to the report, people have been killed in shrimp farm–related violent conflict in at least eleven countries including Indonesia, India, Brazil, Bangladesh, Honduras, Ecuador, the Philippines, Mexico, Guatemala, Thailand and Vietnam, and child labour has been reported in Bangladesh, Thailand, Indonesia, Sri Lanka, India, Cambodia, Peru, Myanmar (Burma) and Ecuador (Environmental Justice Foundation, 2003). In some cases, government and police are complicit in these human rights abuses, which are often perpetrated without fear of retribution.

An investigative article published by the Associated Press in 2015 found that human trafficking in Thailand was part of the supply chain of shrimp en route to North American markets, and that migrants, including children, were being sold and forced to peel shrimp. At the Gig Peeling Factory, Burmese workers reported working sixteen-hour days peeling shrimp with their hands in freezing cold water for little or no money and not being allowed to leave (Anusonadisai, 2015).

There are also significant health risks associated with tropical farmed shrimp. The crowded conditions of the farms make the shrimp more susceptible to disease and parasites. The farmers deal with this by including antibiotics in the feed and using pesticides and fungicides, some of which are known carcinogens and harmful to human health. A 2015 report by *Consumer Reports* found shrimp contaminated with antibiotics not permitted for

Chef Daniel Cullen, part of the culinary team at Finest At Sea, with a spot prawn. Finest At Sea is a BC company that is owned by fishers and that processes all of their catch on site at their retail outlets, providing traceability from ocean to consumer. JENNIFER GIDORA PHOTO

use in shrimp farming in the US as well as bacteria including *E. coli*, MRSA and *Vibrio* (Consumer Reports, 2015).

Slow Fish, an international network promoting good, clean and fair seafood, recommends avoiding tropical farmed shrimp and enjoying sustainably produced shrimp as a special-occasion food. Here in BC, we are extremely fortunate to have locally available shrimp, including sidestripe shrimp (*Pandalopsis dispar*) and spot prawns (*Pandalus platyceros*), which are seasonally and sustainably fished. Entire festivals exist in several coastal communities in BC to celebrate the spot prawn season. Our local shrimp are more expensive than tropical farmed shrimp coming from thousands of kilometres away, but they exemplify "good, clean and fair."

OPEN-NET PEN SALMON FARMS

"We need to end salmon farming in our open oceans now to protect both wild salmon and Indigenous ways of being from extinction."

—TERRY TEEGEE, REGIONAL CHIEF OF THE BC ASSEMBLY OF FIRST NATIONS (BAKER, 2020)

Open-net pen salmon farming is a global industry that is having devastating effects on ecosystems, and in particular on wild salmon. In a 2016 report by SeaChoice, open-net pen farmed Atlantic salmon comprised the highest volume of unsustainable seafood produced in Canada, an estimated 72 per cent (Govender et al., 2016). Open-net pen salmon farming involves the rearing of salmon fry, typically non-native to the location of the farm, in coastal waters where the salmon are contained in pens made up of nets. The nets enable the free flow of water both into and out of the salmon pens, allowing for excrement, bacteria, viruses, parasites and pesticides to freely enter coastal waters. The salmon on these farms produce huge amounts of excrement, which damages ecosystems through contamination and by causing toxic algal blooms. It is estimated that a farm of two hundred thousand salmon produces the same amount of waste as a city of sixty thousand people (Sandvik, Slettehaug and Ilina, 2018). A typical farm in BC contains an average of four hundred thousand to five hundred thousand salmon. Other marine species that live in close proximity to salmon farms, such as clams and oysters, frequently become too contaminated for human consumption.

The extremely high density of fish on the farms makes the salmon more susceptible to disease. The increased risk is due not only to the close proximity of the fish and the massive amount of excrement in the environment but also because the confined, crowded conditions the salmon endure cause them to release stress hormones, which suppress their immune system. According to fish pathologist Gary Marty, 10 per cent of farmed fish in BC don't make it to market, the majority of which die due to environmental conditions including low oxygen, toxic algae and predation (Smart, 2017). In an attempt to fend off disease, farmed salmon are routinely fed large amounts of antibiotics, which are transferred to humans when the salmon is consumed. In addition, farmed salmon require 1.2 kg of feed to produce 1 kg of meat, and a significant part of the feed for farmed salmon is wild fish (Sandvik et al., 2018).

Canada is the fourth-largest producer of farmed salmon in the world, and farmed salmon are BC's largest agri-food export (Department of Fisheries and Oceans, n.d.). There are currently over one hundred open-net pen salmon farms in BC's coastal waters and more than 90 per cent of the salmon farming industry in BC is owned by three Norwegian companies: Marine Harvest, Cermaq and Grieg Seafood (Findlay, 2018). BC is the last place in the Pacific Northwest to allow open-net pen salmon farms: government bans of open-net pen salmon farms exist in Washington, Oregon, California, Alaska and even parts of northern BC.

Most, if not all, of the salmon farms in BC reside in the traditional territories and fishing grounds of Indigenous peoples, descendants of the first human inhabitants of the expanse of land currently recognized as BC. Archaeological findings show they have lived in the Pacific Northwest for at least ten thousand years. Salmon are woven into every aspect of the life and culture of coastal Indigenous peoples, and many coastal First Nations are dependent upon wild salmon for food, ceremony, cultural identity and livelihood. The issue of salmon farming is complex, as the industry provides many jobs and has brought economic prosperity to several remote communities. However, many First Nations and their leaders oppose the presence of salmon farms on their traditional territories, citing damage to the ecosystem and specifically the threat to wild salmon.

In BC we have seven out of the ten species of Pacific salmon and nine thousand distinct salmon populations (Pacific Salmon Foundation, n.d.). Born in rivers deep in the coastal rainforest and beyond, salmon travel hundreds, and for some species thousands, of kilometres across the ocean and then return to their natal river to spawn and die. Salmon *literally* connect the land to the sea in the Pacific Northwest. They provide an abundance of nutrients to countless iconic West Coast species, including orcas, bears, bald eagles and even the canopy-forming trees of the ancient temperate rainforest. Perhaps the most urgent and detrimental effect of open-net pen salmon farms is the risk they pose to wild salmon populations.

Alexandra Morton is a marine biologist who has been studying orcas for decades. In her quest to uncover the cause of the declining numbers of BC's distinct resident orca population, her focus turned toward salmon, the main food source of the resident orcas. During years of research, she discovered that salmon farms were having a detrimental effect on wild salmon populations by spreading disease and parasites. It was Morton who discovered that the Norwegian strain of Piscine orthoreovirus (PRV), a highly contagious virus associated with the fish disease heart and skeletal muscle inflammation (HSMI), was present in salmon farms located directly in the migratory path of the Fraser River population of Chinook salmon. This population of salmon is the main food source (an estimated 82 to 90 per cent) of the critically endangered Southern Resident Killer Whales. Salmon affected by the disease are not strong enough to make the arduous swim upstream in their natal rivers in order to reach their breeding grounds. The disease can also be fatal, especially in juveniles. Morton has been fighting hard to bring this information to the public and to policymakers in order to bring about legislation that will protect wild salmon.

Sea lice are another significant problem associated with salmon farms. Sea lice are a type of external parasite that feed on the mucus, epidermal tissue and blood of salmon and can cause death, especially in juveniles. Sea lice can spread from farms to wild salmon and are able to live off of their host for up to three weeks. The high numbers of confined salmon make salmon farms particularly prone to sea lice infestations, and it is not uncommon for farms to have to cull their own salmon as a result. The DFO's threshold for sea lice on farmed salmon during the migration of wild juvenile salmon, from March 1 until the end of June, is three per fish. The findings of an independent report by Alexandra Morton on migrating juvenile salmon from four areas near salmon farms found high percentages of juvenile salmon infected with sea lice. The highest percentage was found on juvenile salmon migrating past salmon farms near the Discovery Islands off of northeastern Vancouver Island, where 94 per cent of sampled fish were infected with sea lice (Morton, 2020). This area is the migratory path for the Fraser River population of sockeye salmon, which has seen among the largest declines of any BC salmon population. Moreover, salmon farms release large amounts of pesticides into the water in an attempt to suppress sea lice infestations. These pesticides are harmful to the ecosystem and sea lice are becoming increasingly resistant to them.

Open-net pen salmon farms also release huge amounts of "blood water" into the ocean environment. Blood water is the waste from processing the farmed salmon. In 2017 Tavish Campbell, a diver who works in the ecotourism industry, brought the issue of blood water to the public's attention when he captured on film the bloody effluent from two farms off of Vancouver Island. One farm is located in a bay known to be a refuge for migrating juvenile sockeye salmon from strong tidal currents. Blood water samples have since been tested by the Atlantic Veterinary College and were found to contain high levels of PRV, the highly contagious virus associated with the potentially fatal fish disease HSMI.

The escape of farmed Atlantic salmon also has an impact on both the marine ecosystem and river ecosystems. Atlantic salmon are not native to the North Pacific and in addition to being voracious eaters, competing for the food sources of wild salmon and other wild species, may cause other unknown harms. In August of 2017, the nets of a US-owned salmon farm near Victoria broke, and an estimated 305,000 Atlantic salmon escaped into the ecosystem. Farmed Atlantic salmon have been witnessed in the ocean hundreds of kilometres away from any farms. They have also been found in rivers, including the Fraser River, which is a spawning ground for Chinook salmon and contains the largest population of sockeye salmon in BC.

In addition to the distressing ethical issue of confining a fish that in the wild would travel hundreds, and for some species thousands, of kilometres over the course of its life, inhabiting both fresh water and salt water, other significant ethical issues exist. The presence of thousands of fish in one area attracts large predators, including marine mammals. Even though it is illegal to shoot marine mammals in BC, the Canadian government allows salmon farmers to shoot "nuisance" seals and sea lions. Moreover, animals including dolphins, porpoises, seals, sea lions and birds get entangled in the farms' nets and drown.

Farmed salmon can also contain higher levels of organic contaminants such as PCBs and heavy metals such as mercury than wild salmon (Hites et al., 2004; Easton, Luszniak and Von der Geest, 2002). Farmed salmon often contains high levels of antibiotics, which can affect human health both directly by consuming them and indirectly by creating strains of antibiotic-resistant bacteria. Furthermore, while salmon is considered to be an extremely healthy and nutritious food that is high in omega-3s and protein and low in saturated fat, there are differences between wild salmon and farmed salmon: farmed salmon are fed an unnatural diet that includes terrestrial-based foods such as soybeans, wheat and feathers, and have an altered nutritional makeup. For example, farmed salmon have been found to be higher in

Alexandra Morton, independent biologist and activist.

ANISSA REED PHOTO

Seaweeds are among the fastest-growing organisms on the planet.

saturated fat, calories and omega-6 fatty acids than wild salmon. Farmed salmon are also artificially coloured with a synthetic form of astaxanthin in order to appear pink like their wild counterparts.

Sadly, many people choose farmed salmon specifically to help protect wild salmon. The DFO acknowledges the dangers of open-net pen salmon farms, but despite promises to remove them, the farms remain. The latest claim made by the DFO is that by 2025 they will transition to closed-containment farms, located on land. While this would protect wild salmon, it doesn't resolve the ethical issue of the treatment of salmon on the farms, nor the overwhelming increase in energy and water required to operate closed-containment farms. BC wild salmon stocks are currently the lowest in recorded history, and in 2019 there were unprecedented salmon fishery closures, including for recreational fishing.

CHOOSING SUSTAINABLY SOURCED SEAFOOD

It is much more sustainable to enjoy seafood by eating diverse seafoods in season that are selectively and locally caught by sustainable fishing methods that do not damage the ecosystem. Avoiding unsustainable seafoods such as open-net pen farmed Atlantic salmon and most tropical farmed shrimp altogether is a simple yet powerful way to promote more sustainably produced seafood. Ask where your seafood comes from and find a fishmonger that you trust. If you can, buy direct from the fisher or harvester.

SEAWEED AS NOURISHMENT

Seaweeds are one of the most nutrient-dense foods in the world, with an abundance of vitamins, minerals, protein, fibre, essential fatty acids and other nutrients. In a world with rapidly depleting arable land, healthy soil and fresh water, seaweeds are an important food source and have great potential to nourish many people. Seaweed grows prolifically in the wild in many areas and can be farmed in the ocean without the need for fresh water or fertilizers. Seaweeds are incredibly regenerative and are among the fastest-growing organisms on the planet. If cut in the proper place, they can continue to grow after harvesting, and individual seaweeds can be harvested from multiple times in a season. Moreover, seaweed is nourishing to the environment, adding structure, oxygen and primary production to their ecosystem, lowering the acidity of their surrounding waters and sequestering large amounts of carbon. Algae-based ecosystems

Seaweeds are packed with vitamins, minerals and other health-promoting nutrients.

Tips for Eating Sustainably

- Know where your food comes from and how it is produced
- Know your farmer
- Know your fisher and fishmonger
- Avoid unsustainable and unethical foods altogether
- Eat locally
- Eat seasonally
- Eat more foods lower on the food chain

such as kelp forests and coral reefs are essential for supporting healthy fish populations.

I am exceptionally pleased that here in BC, commercial seaweed harvesters are required to hand-cut each individual seaweed in a place that allows it to regenerate. I harvest all of my seaweeds by pruning, leaving some of the seaweed attached and cutting well above the stipe. For the past nineteen seasons, I have harvested all of my *Alaria* from the same small area, approximately 2,500 square metres. I have harvested my bull kelp from the same kelp forest in front of my home. The seaweeds where I harvest remain as lush and abundant today as they did almost two decades ago. If someone were to come along and rip up those same seaweeds, they would quickly disappear. If harvested with care and knowledge, however, seaweeds have the potential to be an enduring source of sustainable, healthy food.

Seaweed farming in Canada and the US is a new and fast-growing industry. Compared with other food sources, seaweeds have the potential to be farmed much more sustainably, requiring no input of fresh water or nutrients. However, like any type of agriculture, there are environmental risks associated with it, and protecting wild seaweed ecosystems is critical.

IN SUMMARY

Our food choices have deep and far-reaching implications for the health of our planet and humanity. Choosing foods that have been sustainably and justly produced is vital to healthy ecosystems and healthy societies. Supporting your local farmer's market and fish market is a wonderful way to eat sustainably. Most often, you are purchasing food that is locally grown, foraged or raised, in season, directly from the food producer. When you have a personal relationship with the land and the people producing your food, trust is built as well as an expectation of integrity.

With my harvest bag full, I take a moment for contemplation before swimming back to shore. CHRIS ADAIR PHOTO

THE LAST HARVEST

Every season there are a handful of days when the visibility is only about two and a half to three feet. Today, September 2, 2014, is one of those days. The ocean is so plump with plankton and organic detritus that it looks like it is snowing under water, until you see close up the stringy, odd shapes of the floating primary source that feeds a whole ocean.

When I stretch my arm out in front of me, my gloved hand appears foggy and hazy. Talk about spooky. No bottom, no top, nothing on any side but that white-green colour of glaciers. Unlike on land, where things far away appear blurry and then gradually sharpen as you approach, objects under water are not there and then suddenly all there in an instant.

Every Pacific Northwest waterman or waterwoman has a love-hate relationship with the frigid waters of our locale. Upon entering the ocean on the west coast of Vancouver Island, any time of year, even in a full 5/4 wetsuit, you have the sensation of slipping into liquid ice. Sometimes the piercing cold makes me feel invincible, like the strongest person in the world. Other times, it beats me straight down to the ground. For surfers, we know that it keeps away the crowds. We whine when there are twenty surfers at the same point break, and then stare in disbelief at a clip from Southern California where there are twenty surfers dropping in on the same wave and dozens of people wait in the take-off zone like a flock of seagulls. Surely bath-temperature water and big sun would draw out more folks than the current ones, who must be hardcore on some level just to enter the water in the first place. Sometimes there is also the added contemplation and stoke required to put your soaking wet wetsuit back on in zero-degree weather, because the surf is still pumping, it's sunset and there is no swell on the horizon.

Out in the kelp forest, I am trying to harvest, but I can see only the top two feet of the fronds as it is slack tide, no current, and they are hanging straight down into the abyss. The ones that look healthy at the top will hopefully be healthy at the bottom. It's a bit of a blind harvest.

This is where it gets fun. I hold my harvest bag closed and dive down into the vertigo-inducing whiteness.

This is where it gets fun. I hold my harvest bag closed and dive down into the vertigo-inducing whiteness. At a certain depth, the pure white nothingness flashes in an instant to a garden of red, olive green and golden brown seaweeds and the occasional fish darting away. I peer around, looking for lingcods and kelp greenlings, then start my ascent. I go from this thriving, colourful underwater seascape back up into the whiteness.

Today is a day when the cold doesn't bite me, and I feel at one in the emerald sea. It is the first real fall-feeling day of rain after a scorching summer of nothing but sun, fire bans, drought and more sun. The Pacific Northwest is dressed in her true colours today: tones of slate grey and a persistent rain that locals know will be here until spring. The rain has driven the tourists away from the beach and the only souls around are me, as I bob weightlessly in the water, and one flying seagull, singing his yearning cry to the empty landscape.

There is a shorepound today that I know is not caused by surface wind waves, but by real groundswell, the kind I have not seen since April. Though small, I can see swell moving through the kelp forests, two hundred metres out to sea. As I step into the ocean, there is an instant drop-off, created by the pounding swell throughout the night. Even as far out as the kelp forests, I can see the surge of the swell, deep enough to engage the seaweeds covering the bottom at twenty-five feet in her dance of push and pull. The sound of the swell pulling back on the wave-smoothed rocks of French Beach is all-consuming. As a lover of music, I find this sound deeply inspiring and its sheer force, filling my entire soundscape almost deafeningly, is awesome.

The last day of a harvest season is always bittersweet. Though I know I should feel celebratory after an abundant harvest season, I feel a sad longing in my heart as I watch the kelp forest disappear into the green. This season will forever be remembered as "the year of the octopus." In addition to the giant Pacific octopus I saw while harvesting bull kelp, I also saw two Pacific red octopuses swimming by me while I harvested *Alaria*.

I trudge back to the shop, pushing my wheelbarrow, looking longingly at the floating kelp behind me, and prepare to hang the last harvest of the season.

As I hang the last piece of bull kelp from cedar racks, I feel satisfied. I exhale. I did it. Again. I shut off the lights in the shop and head out to the woodshed to get my surfboard. The surf season is just beginning.

LAST WORDS

"The ocean stirs the heart, inspires the imagination and brings eternal joy to the soul."

—ROBERT WYLAND

My intent is to inspire a connection between people and seaweed: to ignite wonder upon seeing a flash of emerald green at the beach, mystery at the sight of a large floating kelp the size of a tree, and magic at the sight of a shimmery, iridescent seaweed. I hope this book has opened not only your mind to the wonderful power and beauty of seaweeds, but also your heart.

BC artist Christina Smith created this original painting for me in trade for my hand-harvested wild seaweed. She named it "Sister of the Ocean Realm." CHRISTINA SMITH PAINTING

SEAWEED RECIPES

"Your product is sublime & funky."

—LOUIS CHAREST,
EXECUTIVE CHEF AT GOVERNMENT HOUSE IN OTTAWA, ONTARIO

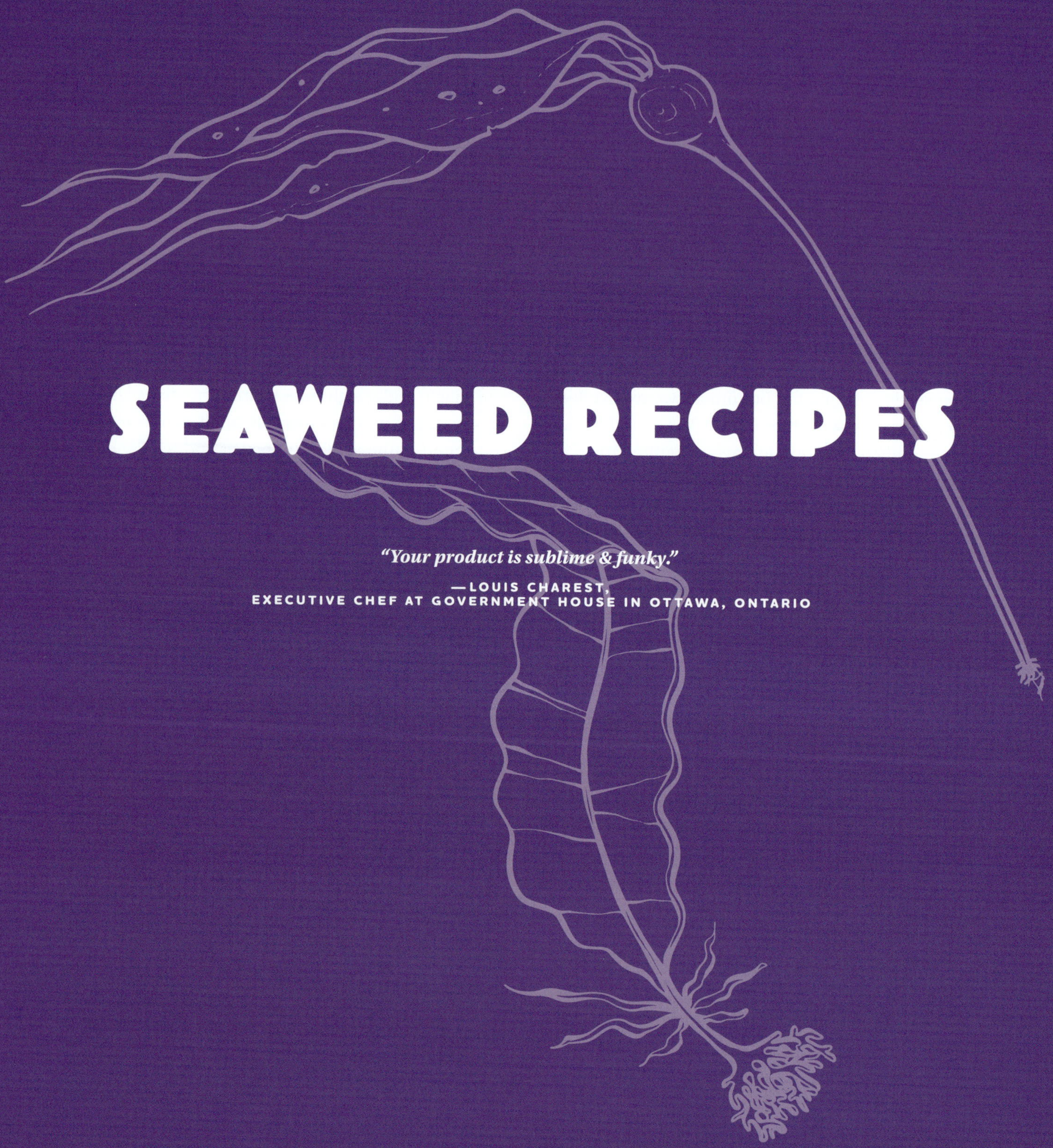

Introduction

"I want to buy your seaweed because I love your story."

—CHEF ATTENDEE AT THE 2018 EVERY CHEF NEEDS A FARMER, EVERY FARMER NEEDS A CHEF EVENT

For the first eleven or twelve years after I started my business, Dakini Tidal Wilds, I sold my seaweed to health food stores and only one restaurant, which served raw, vegan cuisine. My clientele at that time was primarily people wanting to enjoy the tremendous health benefits of seaweed. While I was grateful that seaweed was becoming locally recognized as a potent ally for health, I also felt like a mother who longs for her child to fully unfold their wings, and I wished for seaweed to also be embraced as a culinary treasure.

Seaweed is an unfamiliar food in the West, and unfortunately it had developed a reputation as being "strange and fishy-tasting." Conversely, on the other side of the pond in Asia and especially Japan, Korea and China, where dietary seaweed intake is the highest in the world, diverse varieties are revered as food and the highest quality seaweeds can sell for a small fortune. Canadian West Coast cuisine is increasingly influenced by Japanese and other Asian flavours, and I hoped

SeaCuterie by chef Manpreet Sethi of Aura at the Inn at Laurel Point in Victoria: scallop "mortadella," salmon pâté, Dakini Tidal Wilds winged kelp, spot prawn, charcoal-finished local octopus press, BC *uni* mousse, rye crisps, *piment d'Espelette* and crème fraîche. MANPREET SETHI PHOTO

that our local seaweeds—not just the seaweed wrapped around sushi—would one day become welcome by local palates. Around 2014, seaweed's reputation began to shift from weird to wonderful. I watched with pride as seaweed rose from its humble beginnings to culinary stardom. I credit this wonderful achievement to the incredible talent and admirable values of many of today's trailblazing chefs.

In Canada, we now have a whole new generation of chefs who are not only funky and tattooed, but also dedicated to working with local foods. These chefs are highly skilled at carving out the divine beauty in local, seasonally available ingredients, and they seek out unique, intense flavours. There is no more unique flavour on earth than that of seaweed. It is the embodiment of umami, the long-elusive fifth taste.

Concurrently with this shift in perception, movements such as Slow Food and the 100-Mile Diet were educating people about the importance of knowing where your food comes from. These movements and other trends, such as the recent rise in the numbers and popularity of farmer's markets, have highlighted local, seasonal and artisanal foods as a superior choice for quality and nutrition as well as for ethical and environmental considerations. Moreover, many chefs have strongly conveyed to the public that food grown or foraged locally and harvested at its peak is incomparable in flavour.

Chefs are artists who create a story for the senses with food. Ingredients are the colours in their palette. As a producer of one of their ingredients, my story is part of the stories they create. Trust is critical in the relationship between a chef and a food producer: the level of love, care and attention a chef puts into their dishes must be matched by the love, care and attention a food producer puts into the raw ingredients. Passing my seaweed on to chefs who have both the skill and intent to honour its quality and its story so profoundly is thrilling to me.

It is with deep gratitude to the chefs who have helped seaweed become recognized as a valued and desirable ingredient that I introduce you to this section of the book. I hope you enjoy these beautiful and nourishing recipes, generously contributed by some of the most talented chefs in British Columbia and beyond.

Cooking with Sustainably Sourced Seafood

Please ensure that when you buy seafood, it is sustainably sourced. The best way to ensure this is to know who caught or harvested it. If this is not possible for you, finding a fishmonger you trust is another way to ensure the seafood you are buying was caught or harvested sustainably. A good fishmonger will have a personal relationship with the fishers and harvesters they are getting their seafood from and will be able to tell you exactly where the seafood has come from and the method used to catch or harvest it. Another great way to source sustainable seafood is to buy directly from a community supported fishery (CSF). When you support a CSF, fishers and harvesters receive a fair wage and the seafood that ends up on your table is high-quality domestic seasonal seafood caught using sustainable methods.

Salad for Four

RON GREEN PHOTO

CHEF KRIS BARNHOLDEN
Member of Slow Food Chefs' Alliance
Vancouver, BC

YIELD: 4 servings

"This is a fun, summery dish that Amanda and I did in Torino, Italy, for a demonstration at Terra Madre in 2018. It is a lovely side dish to go with any sort of fish or can be eaten as a light lunch. Have it with a glass of Lambrusco to celebrate Slow Food and delicious ingredients."

— **CHEF KRIS BARNHOLDEN**

0.7 oz (20 g) winged kelp
½ cup + 1 Tbsp (140 mL) fresh-pressed flax oil, or substitute olive oil
2 Tbsp + 1 tsp (35 mL) umesu (sour plum vinegar), or substitute rice vinegar or white wine vinegar
7 oz (200 g) soft tofu, cubed
2 plums, sliced
1 cup (250 mL) heirloom cherry tomatoes, halved
3 shiso leaves, torn into small pieces (or substitute basil or mint)
Salt and pepper to taste

Soak the winged kelp in water for at least 1 hour, then drain and cut into bite-sized strips.

Drizzle oil into vinegar while mixing with a hand blender or whisk.

Mix kelp with remaining ingredients, then dress with the oil and vinegar mix. Garnish with any beautiful fresh herbs you have in the garden or fridge!

KRIS BARNHOLDEN PHOTO

KARA MARTYN PHOTO

Vancouver Island Niçoise Salad

CHEF MAT CLARKE
Be Love
Victoria, BC

YIELD: 2 large salads

SALAD

6 new potatoes
12 green beans
1 cup (250 mL) dried winged kelp, soaked until soft
1 shallot
2 heirloom tomatoes
A couple handfuls of local salad greens, torn

DRESSING

1 Tbsp (15 mL) fresh lemon juice
3 Tbsp (45 mL) apple cider vinegar
3 Tbsp (45 mL) capers
3 Tbsp (45 mL) sunflower seeds
1 Tbsp (15 mL) turbinado sugar
3 sun-dried tomatoes
Pinch of salt
Pinch of black pepper
1 cup (250 mL) extra-virgin olive oil
¼ cup (60 mL) dried winged kelp, soaked until soft
¼ cup (60 mL) fresh parsley
⅛ cup (30 mL) fresh dill

OPTIONAL

Dulse flakes
Sprouts of choice
¼ avocado
1 hard-boiled egg
Seared albacore tuna

Create an ice bath by placing ice cubes and cold water in a sink or large bowl.

Place the potatoes in a medium pot of salted cold water and bring to a boil over high heat. Boil at high heat for about 5 minutes, or until you can poke a fork easily through them.

KATRINA RAIN PHOTO

Add the green beans and allow water to return to a boil, then use a slotted spoon to strain the potatoes into a medium bowl with a splash of apple cider vinegar.

Strain the beans and place them in the ice bath.

Prepare the remaining salad ingredients: Thinly slice the kelp and toss it in with the potatoes. Shave the shallot as thin as possible. Cut the tomato in quarters and season with sea salt and black pepper.

For the dressing: Place the lemon juice, apple cider vinegar, capers, sunflower seeds, sugar, sun-dried tomatoes, salt and pepper in a blender and blend until smooth. With the blender running at a low speed, slowly add the olive oil until combined. Add the kelp and herbs and pulse a couple of times, until combined but maintaining some texture.

Toss the greens with dressing to taste. (Save leftover dressing for other uses.) Place remaining ingredients on and around the greens on two plates and garnish with dulse flakes (if using).

CHEF MCDONALD: DAVID MCILVRIDE PHOTO

CHEF GRIMLEY: PHOTO COURTESY THE CHEF

Yamy Yam Steak with Sunflower Pesto, Kale Apple Salad, Almonds and Seaweed

CHEF MURRAY MCDONALD AND CHEF CHRISTOPHER GRIMLEY
The Bear, The Fish, The Root & The Berry at Spirit Ridge Lake Resort, Winery and Sacred Land
Osoyoos, BC

YIELD: 1 serving

YAM STEAK

1 (6 oz/170 g) yam, cut in half lengthwise and peeled
1 shallot, roughly diced
1 clove garlic, crushed
1 tsp (5 mL) oil
Pinch of thyme and rosemary, chopped

SUNFLOWER SEED PESTO

¼ cup (60 mL) sunflower seeds
¼ cup (60 mL) pumpkin seeds
½ cup (125 mL) Grana Padano, grated (or substitute parmesan)
3 cups (750 mL) arugula
1 clove garlic
1 lemon, juiced
1 cup (250 mL) grape seed oil
Salt to taste

SALAD

1 cup (250 mL) kale, torn
Small handful of dried seaweed of choice
½ tsp (2.5 mL) nutritional yeast
1 Tbsp (15 mL) toasted almonds
1 Tbsp (15 mL) lemon juice
1 Tbsp (15 mL) grape seed oil
½ apple, sliced
5–6 pieces dried apple (or other dried fruit or berries)
Salt to taste

COOK YAM STEAK

Place the peeled yam in a vacuum pack bag with remaining Yam Steak ingredients. Seal then sous vide at 192F (89C) for 1.5 hours, or until tender to the touch.

DAVID MCILVRIDE PHOTO

Alternatively, place the yam in a shallow baking dish with the remaining ingredients for about 1 hour. Remove yam from marinade, place on a parchment-lined baking sheet and bake in a preheated oven at 200F (95C) for 25–30 minutes, or until golden brown and cooked through. While the yam cooks, prepare the pesto and salad.

PREPARE SUNFLOWER SEED PESTO

Pulse all pesto ingredients except salt in a food processer until mostly smooth but with some texture remaining. Stir in salt to taste.

TOSS SALAD

Combine all ingredients except dried apple in a medium mixing bowl. Let sit for 5 minutes to combine the flavours. Garnish with dried apple.

Once the yam is cooked, slice and season with salt to taste, then dress with pesto. Serve with salad on the side.

ANIS ASSARI PHOTO

Dakini Togarashi Steamed Bowl

CHEF LOUIS CHAREST
Government House
Ottawa, Ontario

YIELD: 1 hearty serving

"I started cooking with steam a few years ago when I wanted to really keep track of my caloric intake and get in shape. Cooking with steam and keeping those cooking juices adds to the flavour of the dish and also retains all of the nutritional value of your ingredients. The oil has to be of the highest quality because that's what your body needs, but in this case it's optional since the nuts and seeds already offer quality fats. By the way, the olive oil used in this recipe is from the folks at The Olive Farm on Salt Spring Island. The sea salt is from Salt Spring Sea Salt. I use sea salt to finish my dishes and in my spice mixes, such as the Kelp Spice Mix. I think this recipe is very attainable for most cooks and should be of interest for all levels. The two spice mixtures can be used in a million ways."

— CHEF LOUIS CHAREST

LAURA FERREIRA DA ROSA AND MAYA CHAREST PHOTO

KELP SPICE MIX

0.7 oz (20 g) Salt Spring Island extra-virgin olive oil
1.8 oz (50 g) minced garlic cloves
0.7 oz (20 g) chopped shallots
0.18 oz (5 g) fennel seeds
0.18 oz (5 g) mustard seeds
0.18 oz (5 g) peppercorns
0.7 oz (20 g) paprika
0.18 oz (5 g) dried oregano
0.18 oz (5 g) dried thyme
0.7 oz (20 g) dried winged and bull kelp, crumbled
0.07 oz (2 g) ground cloves
0.18 oz (5 g) ground star anise
0.7 oz (20 g) Salt Spring Island sea salt

DAKINI TOGARASHI

0.4 oz (10 g) ground dried kelp and dulse (see box, page 194)
0.4 oz (10 g) cumin
0.7 oz (20 g) Korean chili flakes
0.18 oz (5 g) ground chilis
0.7 oz (20 g) sesame seeds
0.18 oz (5 g) ground ginger
0.18 oz (5 g) sumac

STEAMED BOWL

5.3 oz (150 g) grains and legumes of your choice
5.3 oz (150 g) reserved cooking liquid
1.8 oz (50 g) dried kelp
7 oz (200 g) mixed raw vegetables of your choice
1 oz (30 g) onion, diced
3.5 oz (100 g) tomato, diced
0.4 oz (10 g) Kelp Spice Mix (see below)
1.8 oz (50 g) nuts and seeds
0.4 oz (10 g) Dakini Togarashi (see next page)
0.18 oz (5 g) Salt Spring Island extra-virgin olive oil

LAURA FERREIRA DA ROSA AND MAYA CHAREST PHOTO

PREPARE KELP SPICE MIX

Preheat oven to its lowest setting (170F/80C or less). In a small pan over low heat, sauté the garlic and shallot in the olive oil. Add the rest of the ingredients and stir to form a paste.

Spread the paste on a parchment paper–lined sheet pan. Dehydrate in the oven for 8 hours. Once

continues...

cooled, coarsely grind the mixture with a mortar and pestle, food processor or coffee grinder (if the paste is too wet to grind, return to the oven until completely dry). Store in an airtight container and use within 3 weeks. The spice mix can be used on potatoes, meats, vegetables, pasta, eggs and more.

PREPARE DAKINI TOGARASHI

Pulse ingredients in grinder or high-speed blender. Keep in a cool, dry place for up to 3 weeks.

ASSEMBLE DAKINI TOGARASHI STEAMED BOWL

Cook grains and legumes according to package directions. Drain grains and legumes, reserving 1 cup plus 2 tablespoons (280 mL) of the cooking liquid (broth). Add the dried kelp to the broth and allow to soak until softened, then remove kelp and chop roughly. Set kelp and broth aside.

Place kelp, mixed vegetables, onion and tomato in a medium bowl. Add legumes and grains. Toss in Kelp Spice Mix and broth. Place in steamer and cook for 8 to 10 minutes. Mix in nuts and seeds, Dakini Togarashi and olive oil. When I make this for myself, I often add additional protein such as tempeh, tofu or a piece of fish. Enjoy!

How to Grind or Flake Seaweed

Put whole, dried seaweed into a coffee grinder, spice grinder or blender and blend at high speed. For thicker seaweeds, such as winged kelp or kombu, a coffee grinder or spice grinder works best. For thinner seaweeds, such as bull kelp and nori, a blender will work fine. There is a limit to how small winged kelp and other thicker seaweeds will flake, but if you blend bull kelp or other very thin seaweeds too long, they will become a powder. For example, bull kelp only needs a few seconds of blending to flake, whereas winged kelp will need 10 to 25 seconds, depending on how small you want the flakes to be.

Wild Watercress, Kelp and Cauliflower Soup with Crispy Fermented Jerusalem Artichokes

ARTUR JANY PHOTO

LAURA JANY
Wild food forager
Kelowna, BC

YIELD: 4 generous servings

"This is a beautiful, vibrant green spring soup that comes together quickly and is simple to prepare. Wild watercress has a refreshing, clean and peppery flavour that is balanced out by the deeply savoury umami taste of kelp, while cauliflower lends its silky texture to this mineral-rich springtime soup. Prepare the fermented Jerusalem Artichokes at least three days in advance."

— LAURA JANY

LAURA JANY PHOTO

CRISPY FERMENTED JERUSALEM ARTICHOKES

3–4 Jerusalem artichokes (sunchokes)
2 Tbsp (30 mL) pickling spice
6 cloves garlic
1 Tbsp (15 mL) Himalayan sea salt
Garlic salt to taste
Olive oil

WILD WATERCRESS, KELP AND CAULIFLOWER SOUP

1 medium onion
2 medium carrots, peeled
2 stalks celery
1 leek
3 Tbsp (45 mL) unsalted butter
1 cup (250 mL) cauliflower florets
3–5 strips dried bull kelp
3 bunches fresh wild watercress, roughly chopped
Sour cream or coconut milk (optional)
Salt (optional)

continues...

PREPARE FERMENTED JERUSALEM ARTICHOKES

Note: It isn't necessary to ferment the Jerusalem artichokes prior to roasting. It does, however, add a tangy flavour and supports ease of digestion. If you prefer to skip the fermentation process, simply wash the Jerusalem artichokes, then proceed to the roasting steps.

To ferment: Gather enough Jerusalem artichokes to fill a small jar. Wash the tubers but do not peel. Add 2 tablespoons (30 mL) of pickling spice and 6 garlic cloves to a medium jar. Add tubers, leaving an inch (2.5 cm) of space at the top.

Dissolve Himalayan sea salt in 4 cups (1 L) of filtered water, then pour into the jar, ensuring the tubers are covered. Close jar and leave at room temperature until fermentation begins, about 3 to 4 days. When the contents are actively bubbling, store the jar in the fridge until ready to use. (The fermented Jerusalem artichokes will keep for up to 1 month.)

PREPARE WILD WATERCRESS, KELP AND CAULIFLOWER SOUP

Preheat oven to 425F (220C) and line a baking sheet with parchment.

Peel and roughly chop onion, carrots, celery and leek. In a large saucepan, melt butter and add onion, carrots, celery, leek, cauliflower and kelp strips. Stir and cook until onion is translucent, 5 to 6 minutes. Add enough water to cover vegetables and bring to a boil, then reduce to a simmer and cover. Simmer until vegetables have softened, about 15 to 20 minutes.

ROAST THE CRISPY FERMENTED JERUSALEM ARTICHOKES

While the soup cooks, slice raw or fermented Jerusalem artichokes thinly with a mandoline. Use a paper towel to pat the slices dry. Toss slices with garlic salt and a drizzle of olive oil until coated. Roast on the prepared baking sheet until the edges are starting to brown and chips are crispy, about 20 minutes, tossing occasionally to ensure even roasting.

FINISH SOUP AND SERVE

Remove soup from heat, add watercress and cover. Allow the greens to gently wilt, about 2 to 3 minutes. Use an immersion blender to purée soup to a smooth consistency.

Ladle puréed soup into bowls. Stir in a spoonful of sour cream or coconut milk, if using. Add salt to taste if needed (bull kelp often provides enough salt and additional salt may not be necessary). Garnish with Crispy Fermented Jerusalem Artichokes.

Okanagan Sweet Corn and Potato Chowder with Wild Mushrooms and Kelp

CHEF PATRICK GAYLER
The Terrace Restaurant at Mission Hill Family Estate
Kelowna, BC

YIELD: 12 one-cup (250-mL) servings

"Late summer and early fall is a fantastic time to enjoy some really exceptional Okanagan ingredients like sweet corn, wild mushrooms and creamy new potatoes. This simple vegetarian recipe really benefits from the salty, savoury flavour of beautiful winged kelp."

— CHEF PATRICK GAYLER

ROASTED VEGETABLE STOCK

3 corn cobs
1 cup (250 mL) carrot, roughly chopped
1 cup (250 mL) celery, roughly chopped
2 cups (475 mL) onion, roughly chopped
Oil
Salt and pepper
8½ cups (2 L) cold water
2 oz (60 g) dried winged kelp, chopped

CHOWDER

¼ cup + 3 Tbsp (105 mL) butter
½ onion, finely diced
1 clove garlic, minced
1 sprig thyme
9 oz (250 g) wild mushrooms of your choice, cleaned and chopped
⅔ cup + 3 Tbsp (205 mL) apple juice
18 oz (500 g) potatoes, peeled and medium diced
Additional dried winged kelp, crumbled into small flakes, for garnish
Herbs such as chives, tarragon or dill (optional)

PREPARE STOCK

Preheat oven to 350F (180C).

Remove corn kernels from cobs; you should have just over 2 cups (475 mL). Set the kernels aside in a small bowl and save the cobs.

continues...

MICHIEL MEYBOOM PHOTO

Combine the carrot, celery, onion and corn cobs in a roasting pan. Drizzle with oil and sprinkle with a pinch of salt and pepper. Roast until vegetables start to turn golden brown, about 20 minutes.

Once browned, transfer the vegetables to a pot and cover with water. Bring to a light simmer for 30 minutes. Remove from heat.

Stir in kelp, cover with a lid and allow to stand for an hour. Strain out solids and allow stock to cool. The stock is now ready to go!

COOK CHOWDER

In a large sauce pot combine butter, onion and garlic. Sauté over medium-high heat until soft, then add thyme, prepared corn kernels and mushrooms. Season generously with salt and pepper and continue to sauté until the corn and mushrooms are tender and the mixture is beginning to brown slightly.

Now add in the apple juice, 8 cups (2 L) of the prepared stock and the potatoes. Simmer gently with a lid on until potatoes are fork-tender, about 12 minutes, then remove from heat. Adjust seasoning and add flaked kelp to taste (it will be salty). The soup is now ready to go! Garnish with chopped chives, tarragon or dill, if using, and enjoy.

Smoked Whitefish Dip and Winged Kelp Seaweed Crackers

MARC-ANDRÉ LAVOIE PHOTO

CHEF CHUCK HUGHES
Garde Manger and Le Bremner
Montreal, Quebec

YIELD: roughly 36 crackers

"This recipe was inspired by my summer of 2020. Because the restaurants were closed and travel was not allowed due to the pandemic, I was really able to enjoy this time with my boys: Henri, four, and Charlie, six. We spent most of the summer at the cottage, where we love to fish! We were catching a lot of whitefish that summer and this has become a family classic."

— **CHEF CHUCK HUGHES**

BRINE

4 cups (1 L) water
½ cup (125 mL) salt
½ cup (125 mL) brown sugar
1 Tbsp (15 mL) peppercorns
2 garlic cloves, smashed

SMOKED WHITEFISH

2 pieces whitefish (3–4 lbs/1.4–1.8 kg each), cleaned and filleted*

**When preparing the fish, I leave the bones in and the skin on with scales. This keeps it moist during smoking, and I believe it adds flavour.*

CHUCK HUGHES PHOTOS

WHITEFISH DIP

2 cups (475 mL) shredded smoked whitefish
1 cup (250 mL) sour cream or crème fraîche
1 Tbsp (15 mL) mayonnaise
½ tsp (2.5 mL) Dijon mustard
Cracked pepper
Pinch of sea salt
Zest of ½ lemon
1 Tbsp (15 mL) chopped dill

continues...

CHUCK HUGHES PHOTOS

CRACKERS

1 cup (250 mL) flour
¼ cup (60 mL) water
1 tsp (5 mL) maple syrup
¼ cup (60 mL) powdered winged kelp
½ tsp (2.5 mL) garlic powder
½ tsp (2.5 mL) lemon zest
Pinch dried Italian herbs (basil, oregano, rosemary, etc.)
Pinch chili flakes
2 tsp (10 mL) black sesame seeds
2 tsp (10 mL) sesame seeds

BRINING AND SMOKING

Combine brine ingredients in a shallow dish. Place fillets in brine overnight (10 to 12 hours), then pat dry and let dry uncovered in fridge for 3 hours.

Place fish in smoker. Smoke for 4 hours at low temperature. The fish should reach an internal temperature of 160F (71C) to kill any bacteria. When finished, fish will have golden colour and flesh will flake apart with fork.

CRACKERS

Preheat oven to 400F (205C) and line a baking sheet with parchment.

Combine all ingredients except sesame seeds in a medium bowl. Stir well for about 3 minutes, or until fully combined. Wrap up dough and place in fridge for 30 minutes. Remove from fridge and let sit on counter for 10 minutes.

Split dough in half and flatten with a rolling pin. Roll out until dough is about ⅛ inch (0.32 cm) thick. Cut out shapes with a cookie cutter, then spray dough with water and top off with sesame seeds (the water helps the seeds to stick).

Bake in batches on prepared sheets for 8 minutes, or until golden brown and edges are firm. Do not top with salt; the natural salt in the kelp is perfect.

WHITEFISH DIP

Mix dip ingredients together in a medium bowl. Plate and serve.

Scallop "Mortadella"

CHEF MANPREET SETHI
Aura at the Inn at Laurel Point
Victoria, BC

YIELD: 2 torchons

4.5 oz (128 g) bacon fat
10.5 oz (300 g) scallop purée
3.5 oz (100 g) scallop muscle and meat, cut into 0.4-inch (1-cm) cubes
½ cup + 1 Tbsp (140 mL) pistachios, toasted
3 tsp (15 mL) winged kelp, blended to powder
1¾ tsp (8.5 mL) kosher salt

Melt bacon fat and allow to cool slightly. Weigh scallops, then blend scallop purée with melted bacon fat in a blender or food processor. Once paste forms, transfer to a stand mixer, then beat on medium speed with whisk attachment for 2 minutes, or until smooth. Strain the paste through a tamis or sieve, then fold in pistachios, cubed scallop meat and muscle, winged kelp and salt.

Lay out two large pieces of plastic wrap and measure 9 ounces (250 g) of the scallop mixture onto each piece. Roll each half of the mixture into a tightly wrapped log and tie the ends, then seal each log (torchon) in a zipper-lock freezer bag.

Cook torchons in a sous vide water bath at 125F (50C) for 16 minutes. After cooking, immediately place in an ice bath to reduce to room temperature, then chill for 24 hours before serving.

MANPREET SETHI PHOTO

JASON SCHULTZ PHOTO

Albacore Tuna Poke with Bull Kelp, Avocado "Wasabi," Pickled Apple and Crispy Wontons

CHEF DAN BAIN
Lure Restaurant & Bar
Victoria, BC

YIELD: serves 1 as a main course or 2 as an appetizer

PICKLED APPLES

7 Tbsp (105 mL) rice wine vinegar
3 Tbsp + 1 tsp (50 mL) mirin wine
7 Tbsp (105 mL) water
1 granny smith apple, peeled and diced

MARINATED BULL KELP

1 oz (30 g) dried bull kelp
1 tsp (5 mL) sesame oil

SOY GINGER MARINADE

1 lime
$1\frac{1}{4}$ tsp (6 mL) palm sugar
$\frac{1}{3}$ cup (80 mL) grated ginger
$\frac{1}{4}$ cup (60 mL) minced garlic
$\frac{1}{2}$ cup (125 mL) thinly sliced scallions
7 Tbsp (105 mL) mirin wine
Pinch chili flakes
2 tsp (10 mL) sesame oil
1 Tbsp (15 mL) ground dried bull kelp (see box, page 194)

AVOCADO WASABI

2 ripe avocados
1 ripe mango
2 Tbsp + 2 tsp (40 mL) fresh horseradish, peeled and diced
3 Tbsp (45 mL) lime juice
2 Tbsp (30 mL) tamari
2 cups (475 mL) fresh spinach

WONTON KELP CRISPS

4 wonton wrappers, cut into strips
Pinch of dried bull kelp, powdered in a spice grinder
Frying oil

TUNA

5.5 oz (160 g) sashimi-grade albacore tuna, cubed

PICKLE APPLES

Bring vinegar, mirin and water to a boil. Cool and pour over diced apples. Cover and leave on the counter to pickle overnight.

REHYDRATE AND MARINATE BULL KELP

Break dried kelp into small pieces. Bring 2 cups (475 mL) of water to a boil. Once water is boiling, pour it over the kelp to rehydrate and let sit 3 minutes. Once kelp is soft, strain off the water. Cool. Mix the rehydrated kelp with sesame oil and set aside for a minimum of 1 hour.

PREPARE SOY GINGER MARINADE

Zest the lime using a microplane grater, then cut in half and juice the lime. Grate or crumble palm sugar. Place lime juice and zest, sugar and remaining ingredients in a medium bowl and stir until well mixed.

PREPARE AVOCADO "WASABI"

Cut avocados and remove the pits, scoop out flesh and dice. Peel mango, remove fruit from the pit and dice. Combine all ingredients in a blender or food processor and purée until very smooth.

FRY WONTON CRISPS

Fry wonton strips in hot oil until crispy and golden. Season immediately with powdered bull kelp.

ASSEMBLE POKE BOWL

Toss tuna with 3 tablespoons + 1 teaspoon (50 mL) of soy ginger marinade to coat evenly. Pile tuna, marinated bull kelp and 2 tablespoons (30 mL) of pickled apple on a plate and top with crispy wontons. Serve avocado "wasabi" on the side. Enjoy!

JASON SCHULTZ PHOTO

Salmon and Prawn Bull Kelp Roulade

CHEF STEVE WALKER-DUNCAN
Camosun Culinary Arts
Victoria, BC

YIELD: 6 to 8 appetizers

SALMON AND PRAWN MOUSSELINE ROULADES

4.5 oz (128 g) salmon (reserve skin and bones)
9 oz (250 g) prawns (reserve shells)
½ tsp (2.5 mL) salt
1 egg white
½ cup (125 mL) whipping cream, chilled
½ tsp (2.5 mL) lemon juice
White pepper to taste
Tabasco sauce to taste
1 oz (30 g) dried bull kelp
Small bunch chives, sliced

SHELLFISH STOCK

Skin and bones from salmon
Shells from prawns
1 celery stalk, chopped
1 carrot, chopped
½ onion, chopped
Spare bits of bull kelp

SEAWEED SALAD

Bull kelp (reserved from roulades)
1 carrot, thinly sliced
1 stalk celery, thinly sliced
¼ red onion, thinly sliced
1 Tbsp (15 mL) rice wine vinegar
1 Tbsp (15 mL) chili oil (or substitute olive oil)
Chive flowers for garnish (optional)

FRIED BULL KELP

Bull kelp (reserved from roulades)
Vegetable oil for frying

PREPARE SALMON AND PRAWN MOUSSELINE

Remove salmon skin and bones and set them aside for the stock. Peel and devein shrimp and set the shells aside with the salmon skin and bones.

Place salmon and half of the prawns (if some were torn while peeling, use those ones) into a food processor with the salt and pulse until smooth. Add egg white and pulse until incorporated. With the processor running, slowly add cream; mixture will thicken as cream is incorporated. Finish with lemon juice, but do not over-mix. Season to taste with pepper and Tabasco and refrigerate until ready to use.

Rehydrate dried kelp in warm water for 20 to 30 minutes. Drain and rinse well, then pat dry. Save 8 to 10 pieces, approximately 2 inches × 2 inches (5 cm × 5 cm) each, for Fried Bull Kelp.

MAKE SHELLFISH STOCK

Place prawn shells and salmon scraps in a small pot with the chopped celery, carrots and onions, along with any small scraps of rehydrated bull kelp. Cover with cold water and simmer for 20 minutes. Strain out solids and return stock to the pot on a low simmer for poaching the roulades. (Note: The shellfish stock can also be cooled and used in other recipes, such as soups or sauces, if desired. In that case, the roulades can be poached in water.)

ASSEMBLE ROULADES

Slice remaining prawns in half lengthwise. Trim four large sheets of kelp into squares that measure approximately 8 inches × 8 inches (20 cm × 20 cm). Save smaller pieces and trimmings for Seaweed Salad (next page).

Place a kelp sheet on a larger piece of plastic wrap (step 1). There should be about 1 to 1½ inches (2.5 to 4 cm) of overhanging wrap at each end of the roulade. Spread the bottom third of the kelp, nearest to you, with a quarter of the Salmon and Prawn Mousseline, leaving a border of about ¾ to 1 inch (2 to 3 cm) on the right and left sides.

Roulade: step 1. STEVE WALKER-DUNCAN PHOTO

Roulade: step 2. STEVE WALKER-DUNCAN PHOTO

Roulade: step 3. STEVE WALKER-DUNCAN PHOTO

continues...

Fried Bull Kelp. STEVE WALKER-DUNCAN PHOTO

Salmon and Prawn Bull Kelp Roulade.
STEVE WALKER-DUNCAN PHOTO

Place a quarter of the sliced prawns on top of mousseline, then sprinkle a quarter of the chives over top, reserving some for garnish (step 2). Carefully roll the sheet of bull kelp around the prawns and mousseline to form a tight log, using the plastic to help roll the kelp (do not roll the plastic into the roulade).

Roll the cylinder away from you to wrap the plastic tightly around the roulade. Seal up the ends, making sure not to trap air in the plastic (step 3). Do not roll the roulade backward and forward, as it will compress the roulade too much.

Repeat with the remaining ingredients to create 4 small roulades. Place the roulades into the simmering shellfish stock, ensuring that each roulade is fully submerged. Simmer gently for 12 minutes, then remove and place into ice water to cool.

MIX SEAWEED SALAD

Finely slice the extra pieces of kelp set aside from making roulade wraps. Combine with equal amounts of thinly sliced carrot, celery and red onion, as well as leftover chives from the roulades. (Other sturdy vegetables may be added or substituted, but delicate greens such as lettuce or spinach do not hold up as well to the dressing.) Dress the salad with rice wine vinegar and chili oil.

FRY BULL KELP

Heat a small saucepan over low to medium heat with 1 teaspoon (5 mL) of vegetable oil. (You can add a splash of the chili oil for extra heat.) One at a time, carefully place reserved 2-inch (5-cm) kelp squares into oil, cook for about 1 minute, then turn over and finish cooking until crispy. If kelp browns too quickly, turn down the heat and add a few drops of oil to cool. Remove crispy kelp and drain on kitchen towel.

SERVE

Serve roulades with salad and crispy bull kelp on the side, garnishing with reserved chives and chive flowers (if using).

Clam Rice Wraps

CHEF SHIRLEY LANG
Spirit Culinary Excursions
Victoria, BC

YIELD: 16 wraps

WILD WOOD SORREL PURÉE

2 cups (475 mL) foraged wild wood sorrel leaves, washed
½ cup (125 mL) olive oil
¼ cup (60 mL) lemon juice
½ tsp (2.5 mL) salt

CLAM MARINADE

½ cup (125 mL) dry sake
3 Tbsp (45 mL) tamari or soy sauce
1 Tbsp (15 mL) mirin wine
1 Tbsp (15 mL) ground ginger
1 Tbsp (15 mL) minced garlic
1 tsp (5 mL) chili seeds or 1 minced red Thai chili
3 Tbsp (45 mL) prepared Wild Wood Sorrel Purée

CLAMS

2 cups (475 mL) clam meat (about 4 cups/1 L whole clams)

WRAPS AND FILLING

¼ cup (60 mL) fresh or reconstituted winged kelp or other seaweed, julienned
2 cups (475 mL) assorted fresh greens
1 cup (250 mL) sliced cucumber
1 cup (250 mL) sliced carrot
½ cup (125 mL) pickled red onion
½ cup (125 mL) pickled wild sea asparagus
16 large rice paper wrappers

DIPPING SAUCE

3 Tbsp (45 mL) tamari or soy sauce
1 clove garlic, minced
¼ cup (60 mL) dry sake
1 Tbsp (15 mL) sesame oil
1 Tbsp (15 mL) chili seeds
1 Tbsp (15 mL) chopped cilantro

continues...

SHIRLEY LANG PHOTO

PREPARE WILD WOOD SORREL PURÉE

Purée all ingredients in food processor. Store in fridge in covered container for up to 2 weeks.

PREPARE CLAM MARINADE

Combine all ingredients in a small bowl and mix well.

COOK CLAMS

If using raw clams in their shell, deshell them before use. Bring 4 cups (1 L) salted water to a boil, then add clam meat and cook for 2 minutes. Drain clams and add directly to the marinade while still hot. Allow to rest for 15 minutes in the marinade.

If using precooked clams, add them directly to marinade and set aside for 20 minutes. Remove the clams from the marinade.

PREPARE DIPPING SAUCE

Mix all ingredients together in a small bowl and set aside.

ASSEMBLE WRAPS

Prepare each of the filling ingredients in a separate bowl. Use your preferred small-batch pickle recipe for the pickled red onion and sea asparagus. Fillings can be substituted according to the season and personal preference.

Soften rice paper wrappers according to package instructions. Working with one wrapper at a time, lay softened wrap on a clean work surface. Layer on fresh greens and seaweed, then spoon 2 tablespoons (30 mL) of drained clams on top. Add about 1 tablespoon (15 mL) each of the cucumber and carrot and half a tablespoon (7.5 mL) each of the red onion and sea asparagus. Take the side of the wrap closest to you and fold over the filling, then do the same with each end, as gently and tightly as you can without breaking the rice wrap.

Continue assembling wraps until all ingredients are used.

Place each wrap separately on serving platter so that the wraps don't stick to one another. Cut each wrap in half and serve with the prepared Dipping Sauce in separate small bowls.

Grilled Octopus with Papas Arrugadas Nero and Bull Kelp Salsa Verde with Pickled Sea Asparagus

CHEF JEFF VAN GEEST
Miradoro at Tinhorn Creek Vineyards
Oliver, BC

YIELD: 4 to 6 servings

OCTOPUS

1 small octopus (approximately 1–2 lbs/0.5–1 kg)
8 cloves garlic
2 Tbsp (30 mL) smoked paprika
1 tsp (5 mL) chili flakes
1 lemon, halved
1 bay leaf
½ cup (125 mL) olive oil

PICKLED SEA ASPARAGUS

1 lb (455 g) fresh sea asparagus
1¼ cups (310 mL) white wine vinegar
2 garlic cloves
2 Tbsp (30 mL) coriander
2 Tbsp (30 mL) fennel seed
½ cup + 2 Tbsp (155 mL) white sugar
¼ cup (60 mL) kosher salt

PAPAS ARRUGADAS NERO

1 lb (455 g) nugget potatoes
1 cup (250 mL) kosher salt
2 Tbsp (30 mL) squid ink

BULL KELP SALSA VERDE

2½ Tbsp (37 mL) sliced shallots
0.2 oz (6 g) anchovies
1 tsp (5 mL) Dijon mustard
2 Tbsp (30 mL) capers
¾ cup (180 mL) parsley
0.4 oz (10 g) bull kelp
¼ cup + 3 Tbsp (105 mL) olive oil
2½ tsp (12 mL) sherry vinegar
Pinch dried red chili flakes

continues…

PHOTO COURTESY LIONEL TRUDEL PHOTOGRAPHY LTD.

COOK OCTOPUS

Place all ingredients in a pot with 1¼ cups (310 mL) water. Cover with a tight-fitting lid. Place over medium heat and allow to come to a simmer. Liquid will come out of the octopus. Cover, reduce heat to low and simmer for 1 hour, or until easily pierced with a knife. Remove the octopus from the liquid and allow to cool.

PICKLE SEA ASPARAGUS

Place sea asparagus in a medium bowl. Combine remaining ingredients in a small saucepan and bring to a boil. Pour boiling liquid over the sea asparagus and allow to cool.

COOK PAPAS ARRUGADAS NERO

In a large saucepan, combine potatoes and salt with 8 cups (2 L) water and bring to boil over high heat. Reduce heat to medium and cook until easily pierced with a fork, about 12 to 15 minutes.

Pour off all but ½ cup (125 mL) of the liquid. Add the squid ink and gently swirl the pot over low heat until the potatoes are covered completely in the squid ink and all of the moisture has evaporated. Set aside in a warm place.

PREPARE BULL KELP SALSA VERDE

Purée all ingredients in a blender until smooth.

SERVE

Cut octopus into quarters with two legs on each portion. Heat a grill or grill pan to high heat.

Place a spoonful of salsa verde on each plate. Place a few potatoes over the salsa verde. Grill the octopus until it has a light char on all sides and place over potatoes. Garnish with pickled sea asparagus.

You can also use clean sea water to cook the potatoes instead of salted fresh water.

Lingcod Baked in Kelp

ROSS DERRICK PHOTO

JON AND ANNE-MARIE CROFTS
Codfathers Seafood Market
Kelowna, BC

YIELD: 1 serving

"Lingcod (*Ophiodon elongatus*) is a predatory fish that, despite its name, is not a true cod. It is, instead, the only member of the genus *Ophiodon*. Muscular and powerful with large heads and somewhat sharp and pointed front teeth, they remind me of bulldogs. Their skin is mottled green-brown with tinges of yellow to blend in with their ocean environment, making them difficult for their prey to see. Occasionally at our market, one or two green ling will show up in a batch. We call these fish 'cobalt ling' as they have bright turquoise iridescence on their skin, and their meat is also tinted with a blue-green hue reminiscent of the colour of the ocean in the kelp forest. Some think this is caused by diet, others by a pigment in their blood, but whatever the cause, we always feel lucky to have received them. We find the meat to be somehow extra sweet, cooking to an even more appealing white colour than the regular ling. When buying lingcod, we always want to buy fish that has been caught either by longline fishing (baited hooks and lines on the bottom of the ocean) or troll fishing (towed lures in the midwater), and not fish that has been trawled in nets. Hook and line fishing has a lower impact on the environment and also means that the fish retains its firmness and is less prone to bruising. For home cooking, we find simple is best, allowing the flavour of the fish to shine through.

"Cooking the fish surrounded in its natural environment of kelp imparts a lovely umami to the fish and to the natural sauce that forms from the butter, garlic and fish juices. The crispy baked kelp also gives the dish a nice contrast of texture. We like to serve it with local asparagus and with morel mushrooms gathered by a friend. "

— **JON AND ANNE-MARIE CROFTS**

1 oz (30 g) dried bull kelp or winged kelp
6 oz (170 g) portion lingcod, skin on
Kelp flakes (or sea salt)
Pinch black pepper
2 Tbsp (30 mL) butter
1 clove garlic

Preheat the oven to 350F (180C).

Rehydrate kelp by covering it with boiling water and leaving it to sit for 3 minutes, or until soft.

continues...

JON CROFTS PHOTO

JON CROFTS PHOTO

Lay out 2 to 3 sheets of rehydrated kelp, enough to completely wrap the fish, on a cutting board. Place the lingcod portion skin side up on the kelp. (The skin has a fat layer underneath that is delicious, and the skin itself is tasty.) Season the fish liberally with kelp flakes and a good pinch of black pepper. Place the butter and clove of garlic on top of the fish. Wrap the fish completely in the kelp to form a parcel, overlapping a little to keep in all of the flavour.

Place on a baking sheet and bake for 14 minutes, or until the kelp is dried out and crispy on top. Remove from the oven and gently unwrap the crispy kelp from around the fish, but leave it on the plate to serve.

West Coast Biscuits

CHEF OLIVER KIENAST
Member of Slow Food Chefs' Alliance
Wild Mountain Food & Drink
Sooke, BC

YIELD: 8 to 10 biscuits

2½ cups (625 mL) flour
2 Tbsp (30 mL) baking powder
½ tsp (2.5 mL) salt
1 tsp (5 mL) sugar
1 Tbsp (15 mL) ground bull kelp or winged kelp (see box, page 194)
½ cup (125 mL) frozen unsalted butter
1 cup (250 mL) + 1 Tbsp (15 mL) buttermilk, divided
2 Tbsp (30 mL) chopped parsley

Preheat oven to 425F (220C).

Sift all dry ingredients into a large bowl and whisk to combine. Grate frozen butter into the bowl with the dry ingredients. Mix the butter gently using a spoon (not your hands, which will melt the butter). Make a well in the middle of your bowl and pour 1 cup (250 mL) of the buttermilk in. Stir until just combined, then gently fold in parsley.

Turn the dough out onto a lightly floured surface. Gently mould into a rectangle with your hands. Fold one side of the dough into the centre, and then do the same with the other side, folding the dough into thirds like a letter. Gently flatten and then repeat folding process. Wrap dough and chill for 20 minutes in the fridge.

Using a rolling pin, roll out gently to ¾ inch (2 cm) thick. Cut 2- to 3-inch (5- to 7.5-cm) rounds using a cookie cutter. Arrange biscuits on a baking sheet so they are just touching. (The baking sheet can be lined with parchment but it is not necessary.) Brush tops with remaining tablespoon (15 mL) buttermilk. Bake for 15 minutes or until golden brown.

BROOKE FADER PHOTO

Pickled Kelp and Cucumber

CHEF ANNA HUNT
Finest At Sea Ocean Products
Victoria and Vancouver, BC

YIELD: one 3-cup (750-mL) jar

"This is a great pantry staple to add to your pickle repertoire. It makes a great accompaniment to salmon or tuna burgers, and you can add it to your salads for an extra kick and crunch or just dive straight into the jar with a fork (we won't tell)."

— **CHEF ANNA HUNT**

ANNA HUNT PHOTO

1 oz (30 g) kelp (any variety)
1 long English cucumber
½ yellow onion
1 cup (250 mL) water
1 cup (250 mL) rice vinegar
2 Tbsp (30 mL) sugar
1½ Tbsp (22 mL) salt
1 Tbsp (15 mL) mustard seeds
1–2 bird's eye chilis

Rehydrate kelp by covering it with cold water and leaving it to sit for 3 to 5 minutes, or until soft, then slice. Slice cucumber on a diagonal into ¼-inch (0.6 cm) slices. Thinly slice onion.

Fill a large pot with water: it should contain enough water to completely submerge your jar. Bring to a boil over high heat.

In a smaller pot, combine the water, vinegar, sugar, salt and mustard seeds and bring to a boil, stirring to allow the salt and sugar to dissolve. Mix kelp, cucumber and onion in a bowl so they are well combined and pack into a clean and sterilized 3-cup (750-mL) jar with the chili pepper(s). Pour boiling vinegar mixture over the jarred ingredients and close the jar, ensuring that the rim of the jar is wiped clean so it seals properly.

Gently lower the jar into the pot of boiling water, ensuring that it is completely submerged. Boil for 10 minutes, then remove and allow to cool. (When cool, press the top of the jar to test the seal; it should feel solid.) Pickles will keep, unopened, for up to a year. Once opened, keep refrigerated for up to 4 weeks.

Chocolate Macadamia Ginger Kelp Pavement

CHRISTINE HOPKINS
From the Garden . . . at French Beach
Victoria, BC

YIELD: about 2.2 lbs (1 kg)

"I created Kelp Pavement to serve at our seaweed workshops because I love dark chocolate and the natural sea salt in kelps. You can use other nuts, but the macadamia nuts, ginger and orange zest are wonderful together."

— **CHRISTINE HOPKINS**

- 1 lb (455 g) semi-sweet or bittersweet dark chocolate
- 1 cup (250 mL) macadamia nuts, lightly toasted
- ½ cup (125 mL) candied ginger
- 2 tsp (10 mL) Dakini Tidal Wilds Mermaid's Shake (raw flaked bull kelp and winged kelp)
- Zest of one organic orange

Line a baking sheet with parchment paper.

Melt chocolate in a bain-marie or double boiler, stirring frequently. Roughly chop toasted macadamia nuts and ginger together.

Add nuts, ginger, kelp and orange zest to melted chocolate. Pour onto prepared tray and spread evenly into a thin layer. Let harden in refrigerator for about 30 minutes. Break into small pieces when set.

Store in a tin lined with parchment paper in the refrigerator or freezer. (It keeps very well if well sealed and wrapped in parchment paper.)

CHRISTINE HOPKINS PHOTO

G&T by the Sea

JASON AND ALAYNE MACISAAC
Sheringham Distillery
Sooke, BC

YIELD: 1 drink

"Searching for local botanicals for our gin, we came across a living mermaid on French Beach named Amanda Swinimer (a.k.a. Mermaid of the Pacific) of Dakini Tidal Wilds. Her hand-harvested sustainable winged kelp (*Alaria marginata*) brings the lure of the ocean to our Seaside Gin. With our original location having a view of the sea and the scent of fresh, salty air, our road being named Seaside Drive and a mermaid living down the street, it was serendipitous to add Seaside Gin to our Coastal Craft Spirits."

— **JASON AND ALAYNE MACISAAC**

KATE WHYTE PHOTO

$1\frac{1}{2}$ oz (44 mL) Seaside Gin
3.5 oz (100 mL) Fever-Tree Indian Tonic Water
Lime wheel
Winged kelp for garnish

Fill rocks glass with ice. Add Seaside Gin. Top with Fever-Tree tonic. Garnish with kelp and a lime wheel. Enjoy!

ACKNOWLEDGEMENTS

Bringing this book from vision to reality has been one of the most thrilling journeys of my life, and one that I could not have done alone. I am deeply grateful to the following people who contributed to my book and supported me on this journey.

Jennifer Swinimer Silvester: I can't thank you enough for the countless hours we spent editing on the phone, and for patiently explaining to me grammatical nuances "that/which" come so naturally to you. Thank you for helping organize the book, for contributing to the About the Author section and for your persistent encouragement.

Dan Swinimer: Thank you for your help in coming up with the title for my book, and supporting that process right up until the moment of "print it!"

Bill and Sue Swinimer: Thank you for giving me so many incredible opportunities to explore my passion for the ocean.

Jesse and Turtle Swinimer: Thank you for being a loving and resourceful sounding board.

Kristiana Powell: Thank you for your enduring moral support.

Chris Adair: Thank you for coming out to my favourite bull kelp forest to take underwater photos. Your photos are stunning and provide the reader a visual glimpse into the mystical realm of the kelp forest.

Jennifer Jellett: Thank you, Jen, for bringing your camera to the beach over the years and capturing some great shots, like the magic of a seemingly typical windy coastal day at my favourite *Alaria* garden.

Emma Geiger: Thank you for bringing along your camera almost every time we harvested during your apprenticeship and photographing the seaweeds, and for contributing your beautiful photos to my book.

Agathe Bernard: Thank you for your keen interest in what I do and for capturing it in such a unique and beautiful way on film.

Kate Woods: Thank you for generously contributing your wonderful photos.

Claire Watson: Thank you for creating the beautiful illustrations for my book.

Vera Rønningen: Thank you for writing the foreword and for sharing your deep knowledge of seaweeds with me over the years.

Tien Bui, Kelly Van Houtte and Liz Glowacki: Thank you for contributing your thoughts on the experience of apprenticing with me.

My chef customers: Thank you for creating such inspiring culinary art with the seaweeds I harvest and for generously contributing your recipes.

Heather Good: Thank you for your gift of mermaid art and for allowing me to share it in my book.

Christina Smith: Thank you for creating and contributing your original art.

Lindsey Love: Thank you for your valuable and generous advice and your encouragement.

Shed Simas: Thank you for the beautiful book design.

Caroline Skelton: Thank you for your skilful and compassionate copy edit.

Anna Comfort O'Keeffe, Nicola Goshulak and Harbour Publishing: Thank you for believing in my vision and for your expertise in polishing my original manuscript.

Thank you to everyone else who generously contributed photos.

AGATHE BERNARD PHOTO

GLOSSARY

ALGAE (singular **ALGA**): a group of eukaryotic and prokaryotic photosynthetic organisms that includes both microalgae and macroalgae (seaweeds)

ALGINATE: a salt of alginic acid

ALGINIC ACID: an insoluble, gelatinous polysaccharide found in brown macroalgae, primarily as a salt

ANTIBIOTIC: a substance that kills or inhibits the growth of bacteria

ANTIOXIDANT: a substance that inhibits oxidation

APOPTOSIS: a normal physiological process in the body that programs the self-destruction of specific cells in order to rid the body of viruses, DNA-damaged cells and cancerous cells, and in order to maintain balance during normal growth and development

BLADE: a flat, leaf-like frond of a macroalga (seaweed)

CANCER STEM-LIKE CELLS (CSCs): tumour cells that possess characteristics of regular stem cells and are often resistant to chemotherapy and radiation treatment

CARCINOGEN: a substance that can cause cancer

CARDIOPROTECTIVE: a quality of a compound or substance that protects the heart and vessels from disease

CAROTENOIDS: pigments in plants and algae responsible for their yellow, orange or red colour

CARRAGEENAN: a group of sulphated polysaccharides found in some red algae and used commercially as a stabilizer and thickener

CETACEAN: a marine mammal in the order Cetacea, which includes whales, dolphins and porpoises

CETOLOGIST: a biologist who specializes in the study of cetaceans

CLOSED-LOOP RECYCLING: a recycling process where a product, once it reaches the end of its life cycle, is recycled into another product, and this process is repeated indefinitely

COMMUNITY SUPPORTED FISHERY (CSF): a values-based business model where fishers sell their catch directly to consumers

CYTOTOXIC: a quality of a compound or substance that is toxic to cells

DISCOID HOLDFAST: a type of holdfast that is disc-shaped

ESTROGEN: a steroid hormone responsible for the development and regulation of the female reproductive system and secondary sex characteristics

EUKARYOTIC CELL: a cell that contains membrane-bound organelles such as nuclei and mitochondria

FATTY ACIDS: the primary component of lipids

FREE RADICAL: an atom that has an unpaired electron, typically highly reactive and capable of damaging healthy cells

FROND: the leaf-like part of a macroalga (seaweed)

FUCOIDAN: a sulphated polysaccharide located in the cell walls of brown algae that exhibits diverse and beneficial biological activities

FUCOSTEROL: a sterol found in brown algae that exhibits beneficial biological activities

FUCOXANTHIN: a carotenoid that is the main accessory pigment of photosynthesis in brown algae and that exhibits beneficial biological activities

GAMETE: a reproductive cell

GAMETOPHYTE: one of two alternating stages in the life cycle of algae and plants; a gametophyte is the sexual phase and is haploid, containing only one set of chromosomes

GUT MICROBIOTA: the sum total of diverse microorganisms that inhabit the gut, including bacteria, viruses, protozoa and fungi

HAPTERA: narrow, finger-like projections that comprise the holdfast of some species of macroalgae

HEPATOPROTECTIVE: a quality of a compound or substance that protects the liver

HIGH INTERTIDAL ZONE: an area of the intertidal zone that is the farthest away from the water and is often exposed to air

HOLDFAST: the part of a seaweed that anchors it to a substrate, such as a rock

HYDROPHOBIC: water-repelling

INTERFERONS: a group of proteins produced and released from cells that signal the presence of viruses and melanoma cells and help the body prevent their replication

INTERTIDAL ZONE: the area at the edge of an ocean or sea that is sometimes exposed to air due to tidal action

IN VITRO: a study performed outside of an organism's natural environment, typically in a Petri dish, test tube or microtiter plates

IN VIVO: a study performed on whole living organisms

KELP: brown macroalgae that belong to the order Laminariales

LAMINARIN: a water-soluble polysaccharide found in some species of brown macroalgae

LOW INTERTIDAL ZONE: an area of the intertidal zone that is only exposed to air at very low tides

MACROALGAE: algae with eukaryotic cells that resemble plants and belong to either kingdom Plantae (red and green macroalgae) or kingdom Chromista (brown macroalgae); informally referred to as seaweeds

MACROPHAGE PHAGOCYTOSIS: the process of an immune cell called a macrophage engulfing a pathogen

MICROALGAE: a group of microscopic photosynthetic organisms, which can have either eukaryotic cells or prokaryotic cells

NEUROPROTECTIVE: a quality of a compound or substance that protects the brain

OMEGA-3: a class of polyunsaturated essential fatty acids that lower LDLs in the blood; found in the highest amounts in algae and cold-water fish

OMEGA-6: a class of polyunsaturated essential fatty acids found primarily in nuts, seeds and vegetable oils

OXIDATION: a chemical reaction whereby a molecule loses one or more electrons, which then makes that molecule reactive

PHLOROTANNINS: an extremely heterogeneous group of phenolic compounds found mostly in brown macroalgae and diatoms (a group of microalgae), but also in some red and green macroalgae

PHYCOLOGIST: a biologist who studies algae

PHYCOLOGY: the study of algae

PHYTOESTROGEN: an estrogen that is naturally occurring in some plants and algae

PHYTOPLANKTON: very small (usually microscopic and single-celled) photosynthetic organisms that drift in salt water or fresh water environments (also called microalgae)

POLYSACCHARIDE: a long-chain carbohydrate composed of monosaccharides

PORPHYRAN: a sulphated polysaccharide found in red algae in the genus *Pyropia*

PROKARYOTIC CELL: a cell that lacks membrane-bound organelles such as nuclei and mitochondria; bacteria and some species of microalgae are the only organisms on earth that have prokaryotic cells

REGIME SHIFT: abrupt and persistent changes in the structure and function of an ecosystem

SEAWEED: an informal term that refers to macroalgae

SELECTIVE CELL APOPTOSIS: programmed cell death by self-destruction in only select cells, whereby damaged or tumour cells are destroyed but healthy cells are left intact

The first seaweed I ever remember seeing is the beautiful, golden-hued *Ascophyllum nodosum* on the shores of eastern Nova Scotia in Lunenburg County, where my dad is from.

SLOW FISH: an international movement and grassroots network promoting the values of good, clean and fair seafood

SLOW FOOD: an international movement and grassroots network promoting the values of good, clean and fair food

SODIUM ALGINATE: the sodium salt of alginic acid, a compound present in brown algae

SPORE PATCH: a conspicuous patch containing spores, which appears on some species of seaweed such as *Nereocystis luetkeana*

SPOROPHYLLS: modified parts of the seaweed frond that are reproductive; they contain sporangia

STIPE: the stem-like part of a macroalga (seaweed)

SUBTIDAL: an area of the ocean that is always covered in water and is never exposed to air due to tidal action

T-CELL: a white blood cell that is part of the immune system; it protects the body from infections and may help to fight cancer

THALASSOTHERAPY: the therapeutic, topical use of seaweeds, sea water and other substances from the ocean

THALLUS: the plant-like body of macroalgae

RECOMMENDED READING

Béliveau, Richard, Denis Gingras. 2007. *Cooking with Foods that Fight Cancer*. McClelland & Stewart.

Cooksley, Valerie. 2018. *Seaweed: Nature's Secret to Balancing Your Metabolism, Preventing Disease, and Revitalizing Body & Mind*. 2nd ed. Floramed Publishing.

Druehl, Louis, Rae Hopkins. 2009. *Exploring Kelp: The Kombu Story*. Canadian Kelp Resources Ltd.

Druehl, Louis, Bridgette Clarkston. 2016. *Pacific Seaweeds: A Guide to Common Seaweeds of the West Coast*. Updated and expanded ed. Harbour Publishing.

Rhatigan, Prannie. 2009. *Irish Seaweed Kitchen: The comprehensive guide to healthy everyday cooking with seaweeds*. Booklink.

Hauling a heavy load of bull kelp onto the beach. AGATHE BERNARD PHOTO

BIBLIOGRAPHY

Chlorophyta: The Green Seaweeds, Rhodophyta: The Red Seaweeds and Phaeophyceae: The Brown Seaweeds

Biello, D. 2013. "How much will tar sands oil add to global warming?" *Scientific American*. January 23, 2013. https://www.scientificamerican.com/article/tar-sands-and-keystone-xl-pipeline-impact-on-global-warming/.

Chavoshi, Ranah. 2018. "The effects of diluted bitumen on the developmental stages of the intertidal brown alga, rockweed (*Fucus distichus*)." Presentation, 5th joint meeting of the Phycological Society of America and International Society of Protistologists. July 30, 2018. Vancouver, BC.

Cornwall, Warren. 2019. "Ocean heat waves like the Pacific's deadly 'blob' could become the new normal." *Science*. January 31, 2019. https://www.sciencemag.org/news/2019/01/ocean-heat-waves-pacific-s-deadly-blob-could-become-new-normal.

Druehl, Louis, Bridgette Clarkston. 2016. *Pacific Seaweeds: A Guide to Common Seaweeds of the West Coast*. Updated and expanded ed. Harbour Publishing.

Ghose, T. 2012. "Most Ocean Species Remain Undiscovered." *Live Science*. November 15, 2012. https://www.livescience.com/24805-undiscovered-marine-species.html.

Iovenko, Chris. 2019. "Sea Star Wasting Disease, Revisited." *Hakai*. January 30, 2019. https://www.hakaimagazine.com/news/sea-star-wasting-disease-revisited/.

Kerlin, Kat. 2019. "California's Crashing Kelp Forest." *UC Davis: Science and Climate*. October 21, 2019. https://climatechange.ucdavis.edu/news/californias-crashing-kelp-forest/.

Lee, Lynn. 2019. "Chiix̱uu Tll iinasdll: A story of people nurturing seafood to grow." Presentation, Shaw Centre for the Salish Sea. December 11, 2019. Victoria, BC.

National Oceanic and Atmospheric Administration. 2018. "How Many Species Live in the Ocean?" National Ocean Service (website). March 1, 2018. https://oceanservice.noaa.gov/facts/ocean-species.html.

Rogers-Bennett, Laura. 2017. "Marine heat wave impacts bull kelp forest ecosystem in Northern California." Presentation, 71st Annual Meeting of the Phycological Society of America. June 6, 2017. Monterey, California.

Schiltroth, Braeden. 2018. "Effects of warm ocean temperatures on bull kelp forests in the Salish Sea." Presentation, 5th joint meeting of the Phycological Society of America and International Society of Protistologists. July 31, 2018. Vancouver, BC.

Seaweeds of Alaska. n.d. Seaweeds of Alaska (website). Accessed November 12, 2018. http://www.seaweedsofalaska.com.

Starzomski, Brian. n.d. Biodiversity of the Central Coast (website). Accessed October 15, 2018. https://www.centralcoastbiodiversity.org.

The Marine Detective. 2019. "Extinction? Every Individual's Name Was Known." The Marine Detective (website). March 6, 2019. https://themarinedetective.com/2019/03/06/extinction-every-individuals-name-was-known/.

Wernberg, T. 2018. "Flattening of global kelp forests: tales of turf, temperature and tropical herbivores." Presentation, 5th joint meeting of the Phycological Society of America and International Society of Protistologists. August 2, 2018. Vancouver, BC.

Wernberg, T., K. Krumhansl, K. Filbee-Dexter, M. Pedersen. 2019. "Status and trends for the world's kelp forests." In *World Seas: An Environmental Evaluation*, 2nd ed., 57–78. Elsevier. http://dx.doi.org/10.1016/B978-0-12-805052-1.00003-6.

Zilliac, Chloe. 2019. "Return of the Blob: Marine Heat Wave Wreaks Havoc in the Pacific." *Sierra*. October 8, 2019. https://www.sierraclub.org/sierra/return-blob-marine-heat-wave-wreaks-havoc-pacific.

Seaweed & Health

Abad, M., L. Bedoya and P. Bermejo. 2008. "Natural marine anti-inflammatory products." *Mini-Reviews in Medicinal Chemistry* 8: 740–54.

Abidov, M., Z. Ramazanov, R. Seifulla, S. Grachev. 2010. "The effects of Xanthigen™ in the weight management of obese premenopausal women with non-alcoholic fatty liver disease and normal liver fat." *Diabetes Obes. Metab.* 12: 72–81. https://dx.doi.org/10.1111/j.1463-1326.2009.01132.x.

Adams, Y., S. Smith, R. Schwartz-Albiez, K. Andrews. 2005. "Carrageenans inhibit the in vitro growth of *Plasmodium falciparum* and cytoadhesion to CD36." *Parasitol. Res.* 97: 290–94.

Adluri, R., L. Zhan, M. Bagchi, N. Maulik, G. Maulik. 2010. "Comparative effects of a novel plant-based calcium supplement with two common calcium salts on proliferation and mineralization in human osteoblast cells." *Mol. Cell. Biochem.* 340: 73–80. http://dx.doi.org/10.1007/s11010-010-0402-0.

Aisa, Y., Y. Miyakawa, T. Nakazato, H. Shibata, K. Saito, Y. Ikeda, M. Kizaki. 2005. "Fucoidan induces cell apoptosis of human HS-sultan cells accompanied by activation of caspase-3 and down-regulation of ERK pathways." *Am. J. Hematol.* 78: 7–14. https://doi.org/10.1002/ajh.20182.

Alekseyenko, T., S. Zhanayeva, A. Venediktova, T. Zvyagintseva, T. Kuznetsova, N. Besednova, T. Korolenko. 2007. "Antitumor and antimetastatic activity of fucoidan, a sulfated polysaccharide isolated from the Okhotsk sea *Fucus evanescens* brown alga." *Bull. Exp. Biol. Med.* 143: 730–32.

Apostolidis, E., C.M. Lee. 2010. "In vitro potential of *Ascophyllum nodosum* phenolic antioxidant-mediated α-glucosidase and α-amylase inhibition." *Journal of Food Science* 75: 97–102.

Aslam, M., I. Bergin, K. Jepsen, J. Kreider, K. Graf, M. Naik, S. Goldstein, J. Varani. 2013. "Preservation of bone structure and function by *Lithothamnion* sp. derived minerals." *Biol. Trace Elem. Res.* 156: 210–20. https://doi.org/10.1007/s12011-013-9820-7

Astrup, A., S. Bügel. 2010. "Micronutrient deficiency in the aetiology of obesity." *Int. J. Obesity* 34: 947–48.

Athukorala, Y., K.-N. Kim, Y.-J. Jeon. 2006. "Antiproliferative and antioxidant properties of an enzymatic hydrolysate from brown alga, *Ecklonia cava*." *Food Chem. Toxicol. Int. J. Publ. Br. Ind. Biol. Res. Assoc.* 44: 1065–74. http://dx.doi.org/10.1016/j.fct.2006.01.011.

Aviello, G., S. Amu, S. Saunders, P. Fallon. 2014. "A mineral extract from red algae ameliorates chronic spontaneous colitis in IL-10 deficient mice in a mouse strain dependent manner." *Phytother. Res.* 28: 300–304. http://dx.doi.org/10.1002/ptr.4989.

Baik, I., M. Lee, N.-R. Jun, J.-Y. Lee, C. Shin. 2013. "A healthy dietary pattern consisting of a variety of food choices is inversely associated with the development of metabolic syndrome." *Nutr. Res. Pract.* 7: 233–41. http://dx.doi.org/10.4162/nrp.2013.7.3.233.

Beppu, F., Y. Niwano, T. Tsukui, M. Hosokawa, K. Miyashita. 2009. "Single and repeated oral dose toxicity of fucoxanthin (FX), a marine carotenoid, in mice." *J. Toxicol. Sci.* 34: 501–10.

Biban, B., C. Lichiardopol. 2017. "Iodine deficiency, still a global problem?" *Curr. Health Sci. J.* 43: 103–11. http://dx.doi.org/10.12865/CHSJ.43.02.01.

Bocanegra, A., S. Bastida, J. Benedí, M. Nus, J. Sánchez-Montero, F. Sánchez-Muniz. 2009. "Effect of seaweed and cholesterol-enriched diets on postprandial lipoproteinaemia in rats." *Br. J. Nutr.* 102: 1728–39. https://doi.org/10.1017/S000711450999105X.

Boopathy, N.S., K. Kathiresan. 2010. "Anticancer drugs from marine flora: An overview." *Journal of Oncology* 2010: 1–18.

Brown, E.S., P.J. Allsopp, P.J. Magee, C.I.R. Gill, S. Nitecki, C.R. Strain, E.M. McSorley. 2014. "Seaweed and human health." *Nutr. Rev.* 72: 205–16. http://dx.doi.org/10.1111/nure.12091.

Brown, J., L. Attardi. 2005. "The role of apoptosis in cancer development and treatment response." *Nature Reviews Cancer* 5: 231–37.

Bu, T., M. Liu, L. Zheng, Y. Guo, X. Lin. 2010. "α-glucosidase inhibition and the in vivo hypoglycemic effect of butyl-isobutyl-phthalate derived from the *Laminaria japonica* rhizoid." *Phytother. Res.* 24: 1588–91.

Buck, C.B., C.D. Thompson, J.N. Roberts, M. Müller, D.R. Lowy, J.T. Schiller. 2006. "Carrageenan is a potent inhibitor of papillomavirus infection." *PLOS Pathogens* 2: 671–80.

Campo, V.L., D.F. Kawano, D.B. da Silva Jr., I. Carvalho. 2009. "Carrageenans: Biological properties, chemical modifications and structural analysis—a review." *Carbohydr. Polym.* 77: 167–80.

Canadian Cancer Society. n.d. "Breast Cancer Statistics." Canadian Cancer Society (website). Accessed March 20, 2020. https://www.cancer.ca/en/cancer-information/cancer-type/breast/statistics/?region=on.

Cardoso, S., O. Pereira, A. Seca, D. Pinto, A. Silva. 2015. "Seaweeds as preventative agents for cardiovascular diseases: From nutrients to functional foods." *Mar. Drugs* 13: 6838–65. http://dx.doi.org/10.3390/md13116838.

Chen, H.-Y., T.-C. Huang, L.-C. Lin, T.-M. Shieh, C.-H. Wu, K.-L. Wang, Y.-H. Hong, S.-M. Hsia. 2018. "Fucoidan inhibits the proliferation of leiomyoma cells and decreases extracellular matrix-associated protein expression." *Cell Physiol. Biochem.* 49: 1970–86. https://dx.doi.org/10.1159/000493660.

Chen, J.-H., J.-D. Lim, E.-H. Sohn, Y.-S. Choi, E.-T. Han. 2009. "Growth-inhibitory effect of a fucoidan from brown seaweed *Undaria pinnatifida* on *Plasmodium* parasites." *Parasitol. Res.* 104 (2): 245–50.

Cho, S., D. Han, S.-B. Kim, M. Yoon, H. Yang, Y.-H. Jin, J. Jo, H. Yong, S.-H. Lee, Y.-J. Jeon, M. Shimizu. 2012. "Depressive effect on the central nervous system and underlying mechanism of the enzymatic extract and its phlorotannin-rich fraction from *Ecklonia cava* edible brown seaweed." *Biosci. Biotechnol. Biochem.* 76: 163–68. http://dx.doi.org/10.1271/bbb.110702.

Cho, T.-M., W.-J. Kim, S.-K. Moon. 2014. "AKT signalling is involved in fucoidan-induced inhibition of growth and migration of human bladder cancer cells." *Food Chem. Toxicol.* 64: 344–52. http://dx.doi.org/10.1016/j.fct.2013.12.009.

Cho, Y.-S., W.-K. Jung, J.-A. Kim, I.-W. Choi, S.-K. Kim. 2009. "Beneficial effects of fucoidan on osteoblastic MG-63 cell differentiation." *Food Chem. Toxicol.* 116: 990–94.

Choi, J.-G., O.-H. Kang, O.-O. Brice, Y.-S. Lee, H.-S. Chae, Y.-C. Oh, D.-H. Sohn, H. Park, H.-G. Choi, S.-G. Kim, D.-W. Shin, D.-Y. Kwon. 2010. "Antibacterial activity of *Ecklonia cava* against methicillin-resistant *Staphylococcus aureus* and *Salmonella* spp." *Foodborne Pathogens and Disease* 7: 435–41.

Chu, W.-L., S.-M. Phang. 2016. "Marine algae as a potential source for anti-obesity agents." *Mar. Drugs* 14(12): 222. http://doi.org/10.3390/md14120222.

Cui, W., Y. Zheng, Q. Zhang, J. Wang, L. Wang, W. Yang, C. Guo, W. Gao, X. Wang, D. Luo, 2014. "Low-molecular-weight fucoidan protects endothelial function and ameliorates basal hypertension in diabetic Goto-Kakizaki rats." *Lab Invest.* 94: 382–93. https://doi.org/10.1038/labinvest.2014.12.

Cui, Y., L. Zhang, T. Zhang, D. Luo, Y. Jia, Z. Guo, Q. Zhang, X. Wang, X.M. Wang. 2010. "Inhibitory effect of fucoidan on nitric oxide production in lipopolysaccharide-activated primary microglia." *Clin. Exp. Pharmacol. Physiol.* 37: 422–28.

Cumashi, A., N.A. Ushakova, M.E. Preobrazhenskaya, A. D'Incecco, A. Piccoli, L. Totani, N. Tinari et al. 2007. "A comparative study of the anti-inflammatory, anticoagulant, antiangiogenic, and antiadhesive activities of nine different fucoidans from brown seaweeds." *Glycobiology* 17: 541–52. http://dx.doi.org/10.1093/glycob/cwm014.

Damonte, E., M. Matulewicz, A. Cerezo. 2004. "Sulfated seaweed polysaccharides as antiviral agents." *Curr. Med. Chem.* 11: 2399–2419. http://dx.doi.org/10.2174/0929867043364504.

Das, S.K., T. Hashimoto, K. Shimizu, T. Yoshida, T. Sakai, Y. Sowa, A. Komoto, K. Kanazawa. 2005. "Fucoxanthin induces cell cycle arrest at G_0/G_1 phase in human colon carcinoma cells through up-regulation of $p21^{WAF1/Cip1}$." *Biochim. Biophys. Acta* 1726(3): 328–35. http://dx.doi.org/10.1016/j.bbagen.2005.09.007.

Das, S.K., R. Ren, T. Hashimoto, K. Kanazawa. 2010. "Fucoxanthin induces apoptosis in osteoclast-like cells differentiated from RAW264.7 cells." *J. Agric. Food Chem.* 58: 6090–95. http://dx.doi.org/10.1021/jf100303k.

Davis, R.E. 1998. "Neurophysiology of glutamatergic signalling and anthelmintic action in *Ascaris suum*: pharmacological evidence for a kainate receptor." *Parasitology* 116: 471–86.

Dawczynski, C., R. Schubert, G. Jahreis. 2007. "Amino acids, fatty acids, and dietary fibre in edible seaweed products."

Food Chem. Toxicol. 103: 891–99. http://dx.doi.org/10.1016/j.foodchem.2006.09.041.

Delma, C.R., S.T. Somasundaram, G.P. Srinivasan, M. Khursheed, M.D. Bashyam, N. Aravindan. 2015. "Fucoidan from *Turbinaria conoides*: a multifaceted 'deliverable' to combat pancreatic cancer progression." *Int. J. Biol. Macromol.* 74: 447–57. http://dx.doi.org/10.1016/j.ijbiomac.2014.12.031.

Dettmar, P., V. Strugala, J. Richardson. 2011. "The key role alginates play in health." *Food Hydrocoll.* 25: 263–66. http://dx.doi.org/10.1016/j.foodhyd.2009.09.009.

Eom, S.H. 2012. "Anti-MRSA (Methicillin-Resistant *Staphylococcus aureus*) Substance Isolated From *Eisenia bicyclis* and Its Action Mechanism." PhD diss., Pukyong National University.

Ewart, V. 2018. "Turning the tide on protein misfolding diseases: Can seaweeds help?" Presentation, 5th joint meeting of the Phycological Society of America and International Society of Protistologists. August 1, 2018. Vancouver, BC.

Fleurence, Joël, Ira Levine. 2016. *Seaweed in Health and Disease Prevention.* Elsevier.

Fujita, T., T. Ohue, Y. Fujii, A. Miyauchi, Y. Takagi. 1996. "Heated oyster shell-seaweed calcium (AAA Ca) on osteoporosis." *Calcif. Tissue Int.* 58: 226–30. http://dx.doi.org/10.1007/BF02508640.

Fujita, T., Y. Fujii, B. Goto, A. Miyauchi, Y. Takagi. 2000. "Peripheral computed tomography (pQCT) detected short-term effect of AAACa (heated oyster shell with heated algal ingredient HAI): A double-blind comparison with $CaCO_3$ and placebo." *J. Bone Miner. Metab.* 18: 212–15. https://doi.org/10.1007/s007740070022.

Fuller, R., J. Cardellina II, Y. Kato, L. Brinen, J. Clardy, K. Snader, M. Boyd. 1992. "A pentahalogenated monoterpene from the red alga *Portieria hornemannii* produces a novel cytotoxicity profile against a diverse panel of human tumor cell lines." *Journal of Medicinal Chemistry* 35 (16): 3007–11.

Funahashi, H., T. Imai, Y. Tanaka, K. Tsukamura, Y. Hayakawa, T. Kikumori, T. Mase et al. 1999. "Wakame seaweed suppresses the proliferation of 7,12-dimethylbenz(a)-anthracene-induced mammary tumors in rats." *Jpn. J. Cancer Res. Gann* 90: 922–27.

Funahashi, H., T. Imai, T. Mase, M. Sekiya, K. Yokoi, H. Hayashi, A. Shibata et al. 2001. "Seaweed prevents breast cancer?" *Jpn. J. Cancer Res. Gann* 92: 483–87.

Gallé, J.-B., B. Attioua, M. Kaiser, A.M. Rusig, A. Lobstein, C. Vonthron-Sénécheau. 2013. "Eleganolone, a diterpene from the French marine alga *Bifurcaria bifurcata* inhibits growth of the human pathogens *Trypanosoma brucei* and *Plasmodium falciparum*." *Mar. Drugs* 11: 599–610.

Ganesan, P., K. Matsubara, T. Sugawara, T. Hirata. 2013. "Marine algal carotenoids inhibit angiogenesis by down-regulating FGF-2-mediated intracellular signals in vascular endothelial cells." *Mol. Cell. Biochem.* 380: 1–9. https://doi.org/10.1007/s11010-013-1651-5.

Gao, Y., C. Li, J. Yin, J. Shen, H. Wang, Y. Wu, H. Jin. 2012. "Fucoidan, a sulfated polysaccharide from brown algae, improves cognitive impairment induced by infusion of Aβ peptide in rats." *Environ. Toxicol. Pharmacol.* 33: 304–11.

Gerber, P., J. Dutcher, E. Adams, J. Sherman. 1958. "Protective effect of seaweed extracts for chicken embryos infected with influenza B or mumps virus." *Proc. Soc. Exp. Biol. Med.* 99 (3): 590–93. http://dx.doi.org/10.3181/00379727-99-24429.

Giffin, J., R. Richards, C. Craft, N. Jahan, C. Leggiadro, T. Chopin, M. Szemerda, S. MacKinnon, K.V. Ewart. 2017. "An extract of the marine alga *Alaria esculenta* modulates α-synuclein folding and amyloid formation." *Neuroscience Letters* 644: 87–93. https://doi.org/10.1016/j.neulet.2017.02.055.

Go, H., H.-J. Hwang, T.-J. Nam. 2010. "A glycoprotein from *Laminaria japonica* induces apoptosis in HT-29 colon cancer cells." *Toxicology in Vitro* 24 (6): 1546–53.

Guo, F., C. Huang, Y. Cui, H. Momma, K. Niu, R. Nagatomi. 2019. "Dietary seaweed intake and depressive symptoms in Japanese adults: A prospective cohort study." *Nutr. J.* 18: 58. http://dx.doi.org/10.1186/s12937-019-0486-7.

Hagiwara, H., K. Wakita, Y. Inada, S. Hirose. 1986. "Fucosterol decreases angiotension converting enzyme levels with reduction of glucocorticoid receptors in endothelial cells." *Biochem. Biophys. Res. Commun.* 139: 348–52.

Han, S.-H., Y.-G. Kim, S.-H. Lee, C.-B. Park, S.-W. Han, H.-J. Jang, H.-J. Lee et al. 2010. "Anti-inflammatory activity of

Codium fragile in macrophages induced by peptidoglycan." *Natural Product Sciences* 16: 153–58.

Hata, Y., K. Nakajima, J.-I. Uchida, H. Hidaka, T. Nakano. 2001. "Clinical effects of brown seaweeds, *Undaria pinnatifida* (wakame), on blood pressure in hypertensive subjects." *J. Clin. Biochem. Nutr.* 30: 43–53. http://dx.doi.org/10.3164/jcbn.30.43.

Heo, S.-J., W.-J. Yoon, K.-N. Kim, G.-N. Ahn, S.-M. Kang, D.-H. Kang, A. Affan, C. Oh, W.-K. Jung, Y.-J. Jeon. 2010. "Evaluation of anti-inflammatory effect of fucoxanthin isolated from brown algae in lipopolysaccharide-stimulated RAW 264.7 macrophages." *Food Chem. Toxicol.* 48: 2045–51.

Holick, C., D. Michaud, R. Stolzenberg-Solomon, S. Mayne, P. Pietinen, P. Taylor, J. Virtamo, D. Albanes. 2002. "Dietary carotenoids, serum beta-carotene, and retinol and risk of lung cancer in the alpha-tocopherol, beta-carotene cohort study." *Am. J. Epidemiol.* 156: 536–47.

Hoshiyama, Y., T. Sekine, T. Sasaba. 1993. "A case-control study of colorectal cancer and its relation to diet, cigarettes, and alcohol consumption in Saitama prefecture, Japan." *Tohoku J. Exp. Med.* 171: 153–65. https://doi.org/10.1620/tjem.171.153.

Hosokawa, M., S. Wanezaki, K. Miyauchi, H. Kurihara, H. Kohno, J. Kawabata, S. Odashima, K. Takahashi. 1999. "Apoptosis-inducing effect of fucoxanthin on human leukemia cell line HL-60." *J-Stage Food Sci. Technol. Res.* 5: 243–46. http://dx.doi.org/10.3136/fstr.5.243.

Hosokawa, M., M. Kudo, H. Maeda, H. Kohno, T. Tanaka, K. Miyashita. 2004. "Fucoxanthin induces apoptosis and enhances the antiproliferative effect of the PPARγ ligand, troglitazone, on colon cancer cells." *Biochim. Biophys. Acta* 1675: 113–19. http://dx.doi.org/10.1016/j.bbagen.2004.08.012.

Hsu, H.-Y., P.-A. Hwang. 2019. "Clinical applications of fucoidan in translational medicine for adjuvant cancer therapy." *Clin. Trans. Med.* 8: 15. https://doi.org/10.1186/s40169-019-0234-9.

Huang, L., K. Wen, X. Gao, Y. Liu. 2010. "Hypolipidemic effect of fucoidan from *Laminaria japonica* in hyperlipidemic rats." *Pharm. Biol.* 48: 422–26. http://dx.doi.org/10.3109/13880200903150435.

Huang, T.-H., Y.-H. Chiu, Y.-L. Chan, Y.-H. Chiu, H. Wang, K.-C. Huang, T.-L. Li, K.-H. Hsu, C.-J. Wu. 2015. "Prophylactic administration of fucoidan represses cancer metastasis by inhibiting vascular endothelial growth factor (VEGF) and matrix metalloproteinases (MMPs) in Lewis tumor-bearing mice." *Mar. Drugs* 13: 1882–1900. http://dx.doi.org/10.3390/md13041882.

Hunter, P. 2012. "The inflammation theory of disease: The growing realization that chronic inflammation is crucial in many diseases opens new avenues for treatment." *EMBO Reports* 13, 968–70. http://dx.doi.org/10.1038/embor.2012.142.

Hwang, H., M. Terada, H.-C. Shin. 2008. "Single dose oral toxicity and 4-weeks repeated oral toxicity studies of *Ecklonia cava* extract." *Seikatsu Eisei* 52: 282–89.

Hwang, P.-A., S.-Y. Chien, Y.-L. Chan, M.-K. Lu, C.-H. Wu, Z.-L. Kong, C.-J. Wu. 2011. "Inhibition of lipopolysaccharide (LPS)-induced inflammatory responses by *Sargassum hemiphyllum* sulfated polysaccharide extract in RAW 264.7 macrophage cells." *Journal of Agricultural and Food Chemistry* 59: 2062–68.

Hwang, P.-A., N.-N. Phan, W.-J. Lu, B.-T. Hieu, Y.-C. Lin. 2016. "Low-molecular-weight fucoidan and high-stability fucoxanthin from brown seaweed exert prebiotics and anti-inflammatory activities in Caco-2 cells." *Food Nutr. Res.* 60: 32033. http://dx.doi.org/10.3402/fnr.v60.32033.

Ii, S.-A., 1999. "The Korean childbirth culture." *Bull. Miyazaki Munic. Univ. Fac. Humanit.* 7: 175–96.

Ikeda, K., A. Kitamura, H. Machida, M. Watanabe, H. Negishi, J. Hiraoka, T. Nakano. 2003. "Effect of *Undaria pinnatifida* (wakame) on the development of cerebrovascular diseases in stroke-prone spontaneously hypertensive rats." *Clin. Exp. Pharmacol. Physiol.* 30: 44–48.

Ishikawa, C., S. Tafuku, T. Kadekaru, S. Sawada, M. Tomita, T. Okudaira, T. Nakazato et al. 2008. "Antiadult T-cell leukemia effects of brown algae fucoxanthin and its deacetylated product, fucoxanthinol." *Int. J. Cancer* 123: 2702–12. http://dx.doi.org/10.1002/ijc.23860.

Itoh, H., H. Noda, H. Amano, C. Zhuaug, T. Mizuno, H. Ito. 1993. "Antitumor activity and immunological properties of marine algal polysaccharides, especially fucoidan, prepared from *Sargassum thunbergii* of Phaeophyceae." *Anticancer Res.* 13: 2045–52.

Itoh, H., H. Noda, H. Amano. 1995. "Immunological analysis of inhibition of lung metastases by fucoidan (GIV-A) prepared from brown seaweed *Sargassum thunbergii*." *Anticancer Res.*, 15 (5B): 1937–47.

Jang, J.-H., J.-H. Lee, H. Chand, J.-S. Lee, Y. Lin, N. Weathington, R. Mallampalli, Y.-J. Jeon, T. Nyunoya. 2016. "APO-9′-fucoxanthinone extracted from *Undariopsis peteseniana* protects oxidative stress-mediated apoptosis in cigarette smoke-exposed human airway epithelial cells." *Mar. Drugs* 14: 140. http://dx.doi.org/10.3390/md14070140.

Jensen, A. 1972. "The nutritive value of seaweed meal for domestic animals." *Proc. Int. Seaweed Symp.* 7: 7–14.

Jensen, M.G., M. Kristensen, A. Astrup. 2012. "Effect of alginate supplementation on weight loss in obese subjects completing a 12-wk energy-restricted diet: A randomized controlled trial." *Am. J. Clin. Nutr.* 96: 5–13. https://dx.doi.org/10.3945/ajcn.111.025312.

Jensen, M.G., M. Kristensen, A. Belza, J. Knudsen, A. Astrup. 2012. "Acute effect of alginate-based preload on satiety feelings, energy intake, and gastric emptying rate in healthy subjects." *Obes. Silver Spring Md.* 20: 1851–58. https://dx.doi.org/10.1038/oby.2011.232.

Jeong, Y.-T., Y. Kim, Y.-M. Jung, D.-C. Park, D.-S. Lee, S.-K. Ku, X. Li et al. 2013. "Low molecular weight fucoidan improves endoplasmic reticulum stress-reduced insulin sensitivity through AMP-activated protein kinase activation in L6 myotubes and restores lipid homeostasis in a mouse model of type 2 diabetes." *Mol. Pharmacol.* 84: 147–57. https://dx.doi.org/10.1124/mol.113.085100.

Jhamandas, J., M. Wie, K. Harris, D. MacTavish, S. Kar. 2005. "Fucoidan inhibits cellular and neurotoxic effects of β-amyloid (Aβ) in rat cholinergic basal forebrain neurons." *Eur. J. Neurol.* 21: 2649–59.

Ji, C.-F., Y.-B. Ji. 2014. "Laminarin-induced apoptosis in human colon cancer LoVo cells." *Oncol. Lett.* 7: 1728–32. http://dx.doi.org/10.3892/ol.2014.1952.

Ji, Y.B., C.F. Ji, H. Zhang. 2012. "Laminarin induces apoptosis of human colon cancer LOVO cells through a mitochondrial pathway." *Molecules* 17: 9947–60. http://dx.doi.org/10.3390/molecules17089947.

Jin, D.-Q., G. Li, J.-S. Kim, C.-S. Yong, J.-A. Kim, K. Huh. 2004. "Preventative effects of *Laminaria japonica* aqueous extracts on the oxidative stress and xanthine oxidase activity in streptozotocin-induced diabetic rat liver." *Biol. Pharm. Bull.* 27: 1037–40. https://doi.org/10.1248/bpb.27.1037.

Jin, D.-Q., C.-S. Lim, J.-Y. Sung, H.-G. Choi, I. Ha, J.-S. Han. 2006. "*Ulva conglobata*, a marine algae, has neuroprotective and anti-inflammatory effects in murine hippocampal and microglial cells." *Neuroscience Letters* 402: 154–58.

Jung, W.-K., Y.-W. Ahn, S.-H. Lee, Y.H. Choi, S.-K. Kim, S.S. Yea, I. Choi et al. 2009. "*Ecklonia cava* ethanolic extracts inhibit lipopolysaccharide-induced cyclooxygenase-2 and inducible nitric oxide synthase expression in BV2 microglia via the MAP kinase and NF-κB pathways." *Food Chem. Toxicol.* 47: 410–17.

Kandale, A., A.K. Meena, M.R. Meda, P. Panda, A.K. Mangal, G. Reddy, R. Babu. 2011. "Marine algae: An introduction, food value and medicinal uses." *J. Pharm. Res.* 4: 219.

Kanegawa, K., H. Harada, H. Myouga, Y. Katakura, S. Shirahata, Y. Kamei. 2000. "Telomerase inhibiting activity in vitro from natural resources, marine algae extracts." *Cytotechnology* 33: 221–27.

Kang, C., Y.B. Jin, H. Lee, M. Cha, E.-T. Sohn, J. Moon, C. Park et al. 2010. "Brown alga *Ecklonia cava* attenuates Type I diabetes by activating AMPK and Akt signaling pathways." *Food Chem. Toxicol.* 48: 509–16.

Kang, S.-M., K.-N. Kim, S.-H. Lee, G. Ahn, S.-H. Cha, A.-D. Kim, X.-D. Yang, M.-C. Kang, Y.-J. Jeon. 2011. "Anti-inflammatory activity of polysaccharide purified from AMG-assistant extract of *Ecklonia cava* in LPS-stimulated RAW 264.7 macrophages." *Carbohydr. Polym.* 85: 80–85.

Kang, S.-M., S.-J. Heo, K.-N. Kim, S.-H. Lee, H.-M. Yang, A.-D. Kim, Y.-J. Jeon. 2012. "Molecular docking studies of a phlorotannin, dieckol isolated from *Ecklonia cava* with tyrosinase inhibitory activity." *Bioorg. Med. Chem.* 20: 311–16. http://dx.doi.org/10.1016/j.bmc.2011.10.078.

Kang, Y., M.N. Kahn, N.H. Park, J.Y. Cho, M.C. Lee, H. Fujii, Y.K. Hong. 2008. "Antipyretic, analgesic, and anti-inflammatory activities of the seaweed *Sargassum fulvellum* and *Sargassum thunbergii* in mice." *J. Ethnopharmacol.* 28: 187–90.

A whole other world exists in the darkness beneath the canopy of a kelp forest. AGATHE BERNARD PHOTO

Kartika, Y., H. Bangun, R. Rosidah. 2018. "Effect of sodium alginate on prevention of hypercholesterolemia and atherosclerosis in rats." *Asian J. Pharm. Clinical Res.* 11: 242–47. http://dx.doi.org/10.22159/ajpcr.2018.v11i6.24768.

Kazłowska, K., T. Hsu, C.-C. Hou, W.-C. Yang, G.-J. Tsai. 2010. "Anti-inflammatory properties of phenolic compounds and crude extract from *Porphyra dentata*." *Journal of Ethnopharmacology* 128: 123–30.

Kellogg, J., M. Grace, M. Lila. 2014. "Phlorotannins from Alaskan seaweed inhibit carbolytic enzyme activity." *Mar. Drugs* 12: 5277–94. http://dx.doi.org/10.3390/md12105277.

Kim, E.J., S.Y. Park, J.-Y. Lee, J.H. Park. 2010. "Fucoidan present in brown algae induces apoptosis of human colon cancer cells." *BMC Gastroenterol.* 10: 96.

Kim, E.-K., S.-Y. Ju. 2019. "Asthma and dietary intake of fish, seaweeds, and fatty acids in Korean adults." *Nutrients* 11: 2187. http://dx.doi.org/10.3390/nu11092187.

Kim, K.-J., O.-H. Lee, H.-H. Lee, B.-Y. Lee. 2010. "A 4-week repeated oral dose toxicity study of fucoidan from the sporophyll of *Undaria pinnatifida* in Sprague-Dawley rats." *Toxicology* 267: 154–58. http://dx.doi.org/10.1016/j.tox.2009.11.007.

Kim, K.-J., K.-Y. Yoon, B.-Y. Lee. 2012. "Fucoidan regulate blood glucose homeostasis in C57BL/KSJ m+/+db and C57BL/KSJ db/db mice." *Fitoterapia* 83: 1105–9. http://dx.doi.org/10.1016/j.fitote.2012.04.027.

Kim, K.-N., S.-J. Heo, S.-M. Kang, G. Ahn, Y.-J. Jeon. 2010. "Fucoxanthin induces apoptosis in human leukemia HL-60 cells through a ROS mediated Bcl-xL pathway." *Toxicol. Vitro Int. J. Publ. Assoc. BIBRA* 24: 1648–54. http://dx.doi.org/10.1016/j.tiv.2010.05.023.

Kim, K.-N., S.-J. Heo, W.-J. Yoon, S.-M. Kang, G. Ahn, T.-H. Yi, Y.-J. Jeon. 2010. "Fucoxanthin inhibits the inflammatory response by suppressing the activation of NF- κB MAPKs in lipopolysaccharide-induced RAW 264.7 macrophages." *Eur. J. Pharmacol.* 649: 369–75.

Kim, K.-N., G. Ahn, S.-J. Heo, S.-M. Kang, M.-C. Kang, H.-M. Yang, D. Kim et al. 2013. "Inhibition of tumor growth *in vitro* and *in vivo* by fucoxanthin against melanoma B16F10 cells." *Environ. Toxicol. Pharmacol.* 35: 39–46. http://dx.doi.org/10.1016/j.etap.2012.10.002.

Kim, M.-M., Q. Ta, E. Mendis, N. Rajapakse, W.-K. Jung, H.-G. Byun, Y.-J. Jeon, S.-K. Kim. 2006. "Phlorotannins in *Ecklonia cava* extract inhibit matrix metalloproteinase activity." *Life Sci.* 79: 1436–43. https://doi.org/10.1016/j.lfs.2006.04.022.

Kim, M.S., J.Y. Kim, W.H. Choi, S.S. Lee. 2008. "Effects of seaweed supplementation on blood glucose concentration, lipid profile, and antioxidant enzyme activities in patients with type 2 diabetes mellitus." *Nutr. Res. Pract.* 2: 62–67. http://dx.doi.org/10.4162/nrp.2008.2.2.62.

Kim, R.-K., Y. Suh, K.-C. Yoo, Y.-H. Cui, E. Hwang, H.-J. Kim, J.-S. Kang, M.-J. Kim Y.Y. Lee, S.-J. Lee. 2015. "Phloroglucinol suppresses metastatic ability of breast cancer cells by inhibition of epithelial-mesenchymal cell transition." *Cancer Sci.* 106: 94–101. http://dx.doi.org/10.1111/cas.12562.

Kim, R.-K., N. Uddin, J.-W. Hyun, C. Kim, Y. Suh, S.-J. Lee. 2015. "Novel anticancer activity of phloroglucinol against breast cancer stem-like cells." *Toxicol. Appl. Pharmacol.* 286 (3): 143–150. http://dx.doi.org/10.1016/j.taap.2015.03.026.

Kim, S., J.-I. Kim, J.-W. Choi, M. Kim, N.Y. Yoon, C.-G. Choi, J.-S. Choi, H.-R. Kim. 2013. "Anti-inflammatory effect of hexane fraction from *Myagropsis myagroides* ethanolic extract in lipopolysaccharide-stimulated BV-2 microglial cells." *J. Pharm. Pharmacol.* 65: 895–906. http://dx.doi.org/10.1111/jphp.12049.

Kim, Se-Kwon, ed. 2013. *Marine Pharmacognosy: Trends and Applications.* Taylor & Francis Group.

Kimura, A., M. Kuramoto. 1974. "Influences of seaweeds on the metabolism of cholesterol and anticoagulant actions of seaweeds." *Tokushima J. Exp. Med.* 21: 79–88.

Konishi, I., M. Hosokawa, T. Sashima, H. Kobayashi, K. Miyashita. 2006. "Halocynthiaxanthin and fucoxanthinol isolated from *Halocynthia roretzi* induce apoptosis in human leukemia, breast and colon cancer cells." *Comp. Biochem. Physiol. Toxicol. Pharmacol.* 142: 53–59. http://dx.doi.org/10.1016/j.cbpc.2005.10.005.

Kotake-Nara, E., M. Kushiro, H. Zhang, T. Sugawara, K. Miyashita, A. Nagao. 2001. "Carotenoids affect proliferation of human prostate cancer cells." *Journal of Nutrition* 131: 3303–6.

Kotake-Nara, E., T. Sugawara, A. Nagao. 2005. "Antiproliferative effect of neoxanthin and fucoxanthin on cultured cells." *Fish. Sci.* 71: 459–61.

Kubanek, J., A.C. Prusak, T.W. Snell, R.A. Giese, C.R. Fairchild, W. Aalbersberg, M.E. Hay. 2006. "Bromophycolides C-I from the Fijian red alga *Callophycus serratus.*" *J. Nat. Prod.* 69 (5): 731–35.

Kumar, S., M. Magnusson, L. Ward, N. Paul, L. Brown. 2015. "Seaweed supplements normalise metabolic, cardiovascular and liver responses in high-carbohydrate, high-fat fed rats." *Mar. Drugs* 13: 788–805. https://dx.doi.org/10.3390/md13020788.

Kwon, M.-J., T.-J. Nam. 2006. "Porphyran induces apoptosis related signal pathway in AGS gastric cancer cell lines." *Life Sci.* 79: 1956–62. http://dx.doi.org/10.1016/j.lfs.2006.06.031.

Kwon, Y.-I., D.A. Vattem, K. Shetty. 2006. "Evaluation of clonal herbs of Lamiaceae species against diabetes and hypertension." *Asia Pac. J. Clin. Nutr.* 15: 107–18.

Lahaye, M. 1991. "Marine algae as sources of fibres: Determination of soluble and insoluble dietary fibre contents in some 'sea vegetables.'" *J. Sci. Food Agric.* 54: 587–94. http://dx.doi.org/10.1002/jsfa.2740540410.

Le, Q.-T., Y. Li, Z.-J. Qian, M.-M. Kim, S.-K. Kim. 2009. "Inhibitory effects of polyphenols isolated from marine alga *Ecklonia cava* on histamine release." *Process Biochem.* 44: 168–76.

Lee, H., J.-S. Kim, E. Kim. 2012. "Fucoidan from seaweed *Fucus vesiculosus* inhibits migration and invasion of human lung cancer cell via PI3K-Akt-mTOR pathways." *PLOS One.* http://dx.doi.org/10.1371/journal.pone.0050624.

Lee, H.-R., H. Do, S.-R. Lee, E.-S. Sohn, S. Pyo, E.-H. Son. 2007. "Effects of fucoidan on neuronal cell proliferation: association with NO production through the iNOS pathway." *Prev. Nutr. Food Sci.* 12: 74–78.

Lee, S.-H., M.-H. Park, S.-J. Heo, S.-M. Kang, S.-C. Ko, J.-S. Han, Y.-J. Jeon. 2010. "Dieckol isolated from *Ecklonia cava* inhibits α-glucosidase and α-amylase in vitro, and alleviates postprandial hyperglycemia in streptozotocin-induced diabetic mice." *Food Chem. Toxicol.* 48: 2633–37. http://dx.doi.org/10.1016/j.fct.2010.06.032.

Lee, S.-H., K.-H. Min, J.-S. Han, D.-H. Lee, D.-B. Park, W.-K. Jung, P.-J. Park, B.-T. Jeon, S.-K. Kim, Y.-J. Jeon. 2012. "Effects of brown alga, *Ecklonia cava* on glucose and lipid metabolism in C57BL/KsJ-db/db mice, a model of type 2 diabetes mellitus." *Food Chem. Toxicol.* 50: 575–82. https://doi.org/10.1016/j.fct.2011.12.032.

Li, Y., S.-H. Lee, Q.-T. Le, M.-M. Kim, S.-K. Kim. 2008. "Anti-allergic effects of phlorotannins on histamine release via binding inhibition between IgE and FcεRI." *Journal of Agriculture and Food Chemistry* 56: 12073–80.

Liu, C.-L., Y.-S. Huang, M. Hosokawa, K. Miyashita, M.-L. Hu. 2009. "Inhibition of proliferation of a hepatoma cell line by fucoxanthin in relation to cell cycle arrest and enhanced gap junctional intercellular communication." *Chem. Biol. Interact.* 182: 165–72. https://doi.org/10.1016/j.cbi.2009.08.017.

Lopes, G., E. Pinto, P. Andrade, P. Valentão. 2013. "Antifungal activity of phlorotannins against dermatophytes and yeasts: Approaches to the mechanism of action and influence on *Candida albicans* virulence factor." *PLOS One* 8 (8): e72203. http://dx.doi.org/10.1371/journal.pone.0072203.

Lynch, M., T. Sweeney, J. Callan, J. O'Sullivan, J. O'Doherty. 2010. "The effect of dietary *Laminaria*-derived laminarin and fucoidan on nutrient digestibility, nitrogen utilisation, intestinal microflora and volatile fatty acid concentrations in pigs." *J. Sci. Food Agric.* 90: 430–37. http://dx.doi.org/10.1002/jsfa.3834.

Mackie, A., A. Macierzanka, K. Aarak, N. Rigby, R. Parker, G. Channell, S. Harding, B. Bajka. 2016. "Sodium alginate decreases the permeability of intestinal mucus." *Food Hydrocoll.* 52, 749–55. http://dx.doi.org/10.1016/j.foodhyd.2015.08.004.

Maeda, H., M. Hosokawa, T. Sashima, K. Miyashita. 2007. "Dietary combination of fucoxanthin and fish oil attenuates the weight gain of white adipose tissue and decreases blood glucose in obese/diabetic KK-A^y mice." *J. Agric. Food Chem.* 55: 7701–6. https://dx.doi.org/10.1021/jf071569n.

Mahomoodally, M., D. Lobine, K. Rengasamy, S. Gowrishankar, D. Tewari, G. Zengin, D.H. Kim, I. Sivanesan. 2019. "Marine algae: A potential resource of anti-HSV molecules." *Processes* 7: 887. http://dx.doi.org/10.3390/pr7120887.

Maruyama, H., J. Nakajima, I. Yamamoto. 1987. "A study on the anticoagulant and fibrinolytic activities of a crude fucoidan from the edible brown seaweed *Laminaria religiosa*, with special reference to its inhibitory effect on the growth of sarcoma-180 ascites cells subcutaneously implanted into mice." *Kitasato Arch. Exp. Med.* 60(3): 105–21.

Maruyama, H., H. Tamauchi, M. Hashimoto, T. Nakano. 2003. "Antitumor activity and immune response of Mekabu fucoidan extracted from sporophyll of Undaria pinnatifida." *In Vivo* 17: 245–49.

Maruyama, H., H. Tamauchi, M. Hashimoto, T. Nakano. 2005. "Suppression of Th2 immune responses by mekabu fucoidan from *Undaria pinnatifida* sporophylls." *Int. Arch. Allergy Immunol.* 137: 289–94.

Mendis, E., S.-K. Kim. 2011. "Present and future prospects of seaweeds in developing functional foods." *Adv. Food Nutr. Res.* 64: 1–15.

Menshova, R.V., S.P. Ermakova, S.D. Anastyuk, V.V. Isakov, Y.V. Dubrovskaya, M.I. Kusaykin, B.-H. Um, T.N. Zvyagintseva. 2014. "Structure, enzymatic transformation and anticancer activity of branched high molecular weight laminarin from brown alga *Eisenia bicyclis*." *Carbohydr. Polym.* 99: 101–9. https://doi.org/10.1016/j.carbpol.2013.08.037.

Miao, H., M. Elkin, E. Aingorn, R. Ishai-Michaeli, C. Stein, I. Vlodavsky. 1999. "Inhibition of heparanase activity and tumor metastasis by laminarin sulfate and synthetic phosphorothioate oligodeoxynucleotides." *Int. J. Cancer* 83: 424–31.

Miyake, Y., K. Tanaka, H. Okubo, S. Sasaki, M. Arakawa. 2014. "Seaweed consumption and prevalence of depressive symptoms during pregnancy in Japan: baseline data from the Kyushu Okinawa Maternal and Child Health Study." *BMC Pregnancy Childbirth* 14: 301.

Mochizuki, S., M. Takahashi, A. Yamamoto. 1995. "Effects of 'kajime' *Ecklonia cava* Kjellman on glucose tolerance in rats." *Nippon Suisan Gakkaishi* 61: 81–84.

Mohamed, S., S. Hashim, H. Rahman. 2012. "Seaweeds: A sustainable functional food for complementary and alternative therapy." *Trends Food Sci. Technol.* 23: 83–96. http://dx.doi.org/10.1016/j.tifs.2011.09.001.

Moon, H.E., M. Islam, B.R. Ahn, S.S. Chowdhury, H.S. Sohn, H.A. Jung, J.S. Choi. 2011. "Protein tyrosine phosphatase 1B and α-glucosidase inhibitory phlorotannins from edible brown algae, *Ecklonia stolonifera* and *Eisenia bicyclis*." *Biosci. Biotechnol. Biochem.* 75: 1472–80.

Moo-Puc, R., D. Robledo, Y. Freile-Pelegrin. 2009. "In vitro cytotoxic and antiproliferative activities of marine macroalgae from Yucatán, Mexico." *Ciencias Marinas* 35 (4): 345–58.

Muradian, K., A. Vaiserman, K.-J. Min, V. Fraifeld. 2015. "Fucoxanthin and lipid metabolism: A minireview." *Nutr. Metab. Cardiovasc. Dis.* 25: 891–97. http://dx.doi.org/10.1016/j.numecd.2015.05.010.

Na, H.-J., P.-D. Moon, S.-G. Ko, H.-J. Lee, H.-A. Jung, S.-H. Hong, Y. Seo et al. 2005. "*Sargassum hemiphyllum* inhibits atopic allergic reaction via the regulation of inflammatory mediators." *J. Pharmacol. Sci.* 97: 219–26. https://doi.org/10.1254/jphs.fp0040326.

Nagumo, T. 1983. "Antitumor polysaccharides from seaweeds." *Japanese Kokai Tokkyo Koho, Japan* 5–8: 174, 329.

Namvar, F., R. Mohamad, J. Baharara, S. Zafar-Balanejad, F. Fargahi, H. Rahman. 2013. "Antioxidant, antiproliferative, and antiangiogenesis effects of polyphenol-rich seaweed (*Sargassum muticum*)." *BioMed. Res. Int.* 2013: 604787. http://dx.doi.org/10.1155/2013/604787.

Nara, T., Y. Kamei, A. Tsubouchi, T. Annoura, K. Hirota, K. Iizumi, Y. Dohmoto, Y. Ono, T. Aoki. 2005. "Inhibitory action of marine algae extracts on the *Trypanosoma cruzi* dihydroorotate dehydrogenase activity and on the protozoan growth in mammalian cells." *Parasitol. Int.* 54: 59–64.

Narasimhan, M., S. Pavithra, V. Krishnan, M. Chandrasekaran. 2013. "In vitro analysis of anti-oxidant, antimicrobial and antiproliferative activity of *Enteromorpha antenna, Enteromorpha linza* and *Gracilaria corticata* extracts." *Jundishapur J. Nat. Pharm. Prod.* 8: 151–59.

Natarajan, S., K. Shanmugiahthevar, P. Kasi. 2009. "Cholinesterase inhibitors from *Sargassum* and *Gracilaria gracilis*: Seaweeds inhabiting South Indian coastal areas (Hare Island, Gulf of Mannar)." *Nat. Prod. Res.* 23: 355–69. http://dx.doi.org/10.1080/14786410802156036.

Nishino, H. 1998. "Cancer prevention by carotenoids." *Mutat. Res.* 402: 159–63.

Nishizawa, Kazutosi. 2002. *Seaweeds Kaiso: Bountiful Harvest from the Seas.* Japan Seaweed Association.

Nitta, I., H. Watase, Y. Tomiie. 1958. "Structure of kainic acid and its isomer, allokainic acid." *Nature* 181: 761–62.

Nomura, T., A. Kikuchi, A. Kubodera, Y. Kawakami. 1997. "Proton-donative antioxidant activity of fucoxanthin with 1,1-diphenyl-2-picrylhydrazyl (DPPH)." *Biochem. Mol. Biol. Int.* 24: 360–70.

Ogara, T., T. Takahashi, H. Yasui, K. Uwai, K. Tokuraku. 2015. "Evaluation of the effects of amyloid β aggregation from seaweed extracts by a microliter-scale high-throughput screening system with a quantum dot nanoprobe." *J. Biosci. Bioeng.* 120: 45–50. http://dx.doi.org/10.1016/j.jbiosc.2014.11.018.

O'Gorman, D., C. Tierney, O. Brennan, F. O'Brian. 2011. "The marine-derived, multi-mineral formula, aquamin, enhances mineralisation of osteoblast cells *in vitro*." *Phytother. Res.* 26: 375–80. http://dx.doi.org/10.1002/ptr.3561.

Oiso, T. 1975. "Incidence of stomach cancer and its relation to dietary habits and nutrition in Japan between 1900 and 1975." *Cancer Res.* 35 (11): 3254–58.

Okubo, H., Y. Miyake, S. Sasaki, K. Murakami, K. Tanaka, W. Fukushima, C. Kiyohara et al. 2012. "Dietary patterns and risk of Parkinson's disease: A case-control study in Japan." *Eur. J. Neurol.* 19: 681–88. http://dx.doi.org/10.1111/j.1468-1331.2011.03600.x.

Okuzumi, J., T. Takahashi, T. Yamane, Y. Kitao, M. Inagake, K. Ohya, H. Nishino, Y. Tanaka. 1993. "Inhibitory effects of fucoxanthin, a natural carotenoid, on N-ethyl-N′-nitro-N-nitrosoguanidine-induced mouse duodenal carcinogenesis." *Cancer Lett.* 68: 159–68.

Orhan, I., B. Sener, T. Atici, R. Brun, R. Perozzo, D. Tasdemir. 2006. "Turkish freshwater and marine macrophyte extracts show in vitro antiprotozoal activity and inhibit FabI, a key enzyme of *Plasmodium falciparum* fatty acid biosynthesis." *Phytomedicine* 13: 388–93.

Osteoporosis Canada. n.d. "About the Disease." Osteoporosis Canada (website). Accessed March 30, 2020. https://osteoporosis.ca/about-the-disease/.

O'Sullivan, L., B. Murphy, P. McLoughlin, P. Duggan, P. Lawlor, H. Hughes, G. Gardiner. 2010. "Prebiotics from marine macroalgae for human and animal health applications." *Mar. Drugs* 8: 2038–64. http://dx.doi.org/10.3390/md8072038.

Ozawa, M., T. Ninomiya, T. Ohara, Y. Doi, K. Uchida, T. Shirota, K. Yonemoto, T. Kitazono, Y. Kiyohara. 2013. "Dietary patterns and risk of dementia in an elderly Japanese population: The Hisayama Study." *Am. J. Clin. Nutr.* 97 (5): 1076–82. http://doi.org/10.3945/ajcn.112.045575.

Paradis, M.-E., P. Couture, B. Lamarche. 2011. "A randomised crossover placebo-controlled trial investigating the effect of brown seaweed (*Ascophyllum nodosum* and *Fucus vesiculosus*) on postchallenge plasma glucose and insulin levels in men and women." *Appl. Physiol. Nutr. Metab. Physiol. Appliquée Nutr. Métabolisme* 36: 913–19. https://doi.org/10.1139/h11-115.

Park, H.-J., T. Takatani, T. Noguchi. 1999. "Screening test of domoic acid in algae and shellfishes around Kagoshima where 'Hanayanagi' *Chondria armata* inhabits." *Bull. Fac. Fish., Nagasaki University* 80: 79–81.

health." *Mar. Drugs* 9: 1806–28. http://dx.doi.org/10.3390/md9101806.

Park, H.S., G.-Y. Kim, T.-J. Nam, N.D. Kim, Y.H. Choi. 2011. "Antiproliferative activity of fucoidan was associated with the induction of apoptosis and autophagy in AGS human gastric cancer cells." *J. Food Sci.* 76: T77–T83. http://dx.doi.org/10.1111/j.1750-3841.2011.02099.x.

Park, H.Y., M.H. Han, C. Park, C.-Y. Jin, G.-Y. Kim, I.-W. Choi, N.D. Kim, T.-J. Nam, T.K. Kwon, Y.H. Choi. 2011. "Anti-inflammatory effects of fucoidan through inhibition of NF-κB, MAPK and Akt activation in lipopolysaccharide-induced BV2 microglia cells." *Food Chem. Toxicol.* 49: 1745–52.

Park, J., M. Yeom, D.-H. Hahm. 2016. "Fucoidan improves serum lipid levels and atherosclerosis through hepatic SREBP-2-mediated regulation." *J. Pharmacol. Sci.* 131: 84–92. https://doi.org/10.1016/j.jphs.2016.03.007.

Park, S.-J., K. Lee, D.-S. Lim, S. Lee. 2012. "The sulfated polysaccharide fucoidan stimulates osteogenic differentiation of human adipose-derived stem cells." *Stem Cells Dev.* 21: 2204–11. http://dx.doi.org/10.1089/scd.2011.0521.

Parkin, D., F. Bray, J. Ferlay, P. Pisani. 2005. "Global cancer statistics, 2002." *A Cancer Journal for Clinicians* 55: 74–108.

Patil, N., V. Le, A. Sligar, L. Mei, D. Chavarria, E. Yang, A. Baker. 2018. "Algal polysaccharides as therapeutic agents for atherosclerosis." *Front Cardiovasc. Med.* 5: 153. http://dx.doi.org/10.3389/fcvm.2018.00153.

Paxman, J., J. Richardson, P. Dettmar, B. Corfe. 2008a. "Alginate reduces the increased uptake of cholesterol and glucose in overweight male subjects: a pilot study." *Nutr. Res.* 28: 501–5. http://dx.doi.org/10.1016/j.nutres.2008.05.008.

Paxman, J., J. Richardson, P. Dettmar, B. Corfe. 2008b. "Daily ingestion of alginate reduces energy intake in free-living subjects." *Appetite* 51: 713–19. http://dx.doi.org/10.1016/j.appet.2008.06.013.

Pelkman, C., J. Navia, A. Miller, R. Pohle. 2007. "Novel, calcium-gelled, alginate-pectin beverage reduced energy intake in nondieting overweight and obese women: Interactions with dietary restraint status." *Am. J. Clin. Nutr.* 86: 1595–602.

Peng, J., J.-P. Yuan, C.-F. Wu, J.-H. Wang. 2011. "Fucoxanthin, a marine carotenoid present in brown seaweeds and diatoms: Metabolism and bioactivities relevant to human health." *Mar. Drugs* 9: 1806–28. http://dx.doi.org/10.3390/md9101806.

Pengzhan, Y., L. Ning, L. Xiguang, Z. Gefei, Z. Quanbin, L. Pengcheng. 2003. "Antihyperlipidemic effects of different molecular weight sulfated polysaccharides from *Ulva pertusa* (Chlorophyta)." *Pharmacol. Res.* 48: 543–49. http://dx.doi.org/10.1016/S1043-6618(03)00215-9.

Potts, B., D. Faulkner, R. Jacobs. 1992. "Phospholipase A2 inhibitors from marine organisms." *J. Nat. Prod.* 55: 1701–17.

Rajapakse, N., S.-K. Kim. 2011. "Chapter 2 – Nutritional and digestive health benefits of seaweed." *Adv. Food Nutr. Res.* 64: 17–28. http://dx.doi.org/10.1016/B978-0-12-387669-0.00002-8.

Ramirez-Higuera, A., L. Queveda-Corona, N. Paniagua-Castro, G. Chamorro-Ceballos, A. Milliar-Garcia, M. Jaramillo-Flores. 2014. "Antioxidant enzymes gene expression and antihypertensive effects of seaweeds *Ulva linza* and *Lessonia trabeculata* in rats fed a high-fat and high-sucrose diet." *J. Appl. Phycol.* 26: 597–605. https://dx.doi.org/10.1007/s10811-013-0134-0.

Ramnani, P., R. Chitarrari, K. Tuohy, J. Grant, S. Hotchkiss, K. Philp, R. Campbell, C. Gill, I. Rowland. 2012. "In vitro fermentation and prebiotic potential of novel low molecular weight polysaccharides derived from agar and alginate seaweeds." *Anaerobe* 18: 1–6. http://dx.doi.org/10.1016/j.anaerobe.2011.08.003.

Ren, R., Y. Azuma, T. Ojima, T. Hashimoto, M. Mizuno, Y. Nishitani, M. Yoshida, T. Azuma, K. Kanazawa. 2013. "Modulation of platelet aggregation-related eicosanoid production by dietary F-fucoidan from brown alga *Laminaria japonica* in human subjects." *Br. J. Nutr.* 110: 880–90. http://dx.doi.org/10.1017/S000711451200606X.

Rosa, G., W. Tavares, P. Sousa, A. Pagès, A. Seca, D. Pinto. 2020. "Seaweed secondary metabolites with beneficial health effects: An overview of successes in in vivo studies and clinical trials." *Mar. Drugs* 18: 8. http://dx.doi.org/10.3390/md18010008.

Rösner, H., W. Möller, S. Groebner, P. Torremante. 2016. "Antiproliferative/cytotoxic effect of molecular iodine, povidone-iodine and Lugol's solution in different human carcinoma cell lines." *Oncol. Lett.* 12: 2159–62. http://dx.doi.org/10.3892/ol.2016.4811.

Rupérez, P. 2002. "Mineral content of edible marine seaweeds." *Food Chem. Toxicol.* 79: 23–26. http://dx.doi.org/10.1016/S0308-8146(02)00171-1.

Russo, A., E. Maisto, L. Romis, G. Celentano. 2016. "Use of a natural compound made of *Ecklonia bicyclis* seaweed, *Tribulus terrestris* and water-soluble chitosan oligosaccharide, in male sexual asthenia with mild or mild-moderate erectile dysfunction and serum testosterone levels at the lower limit of normal." *Health* 8: 1668–78. https://dx.doi.org/10.4236/health.2016.815162.

Shibata, T., K. Nagayama, R. Tanaka, K. Yamaguchi, T. Nakamura. 2003. "Inhibitory effects of brown algal phlorotannins on secretory phospholipase A_2s, lipoxygenases and cyclooxygenases." *J. Appl. Phycol.* 15: 61–66.

Shimazu, T., S. Kuriyama, A. Hozawa, K. Ohmori, Y. Sato, N. Nakaya, Y. Nishino, Y. Tsubono, I. Tsuji. 2007. "Dietary patterns and cardiovascular disease mortality in Japan: A prospective cohort study." *Int. J. Epidemiol.* 36: 600–609.

Shin, T., M. Ahn, J.W. Hyun, S.H. Kim, C. Moon. 2014. "Antioxidant marine algae phlorotannins and radioprotection: A review of experimental evidence." *Acta Histochem* 116: 669–74. http://dx.doi.org/10.1016/j.acthis.2014.03.008.

Skibola, C. 2004. "The effect of *Fucus vesiculosus*, an edible brown seaweed, upon menstrual cycle length and hormonal status in three pre-menopausal women: A case report." *BMC Complementary and Alternative Medicine* 4: 10. https://dx.doi.org/10.1186/1472-6882-4-10.

Soeda, S., S. Sakaguchi, H. Shimeno, A. Nagamatsu. 1992. "Fibrinolytic and anticoagulant activities of highly sulfated fucoidan." *Biochem. Pharm.* 43 (8): 1853–58. http://dx.doi.org/10.1016/0006-2952(92)90721-T.

Son, E.-W., D.-K. Rhee, S. Pyo. 2003. "Antiviral and tumoricidal activities of alginate-stimulated macrophages are mediated by different mechanisms." *Arch. Pharm. Res.* 26: 960–66.

Spavieri, J., A. Allmendinger, M. Kaiser, R. Casey, S. Hingley-Wilson, A. Lalvani, M. Guiry, G. Blunden, D. Tasdemir. 2010. "Antimycobacterial, antiprotozoal and cytotoxic potential of twenty-one brown algae (Phaeophyceae) from British and Irish waters." *Phytother. Res.* 24: 1724–29. https://dx.doi.org/10.1002/ptr.3208.

Spavieri, J., A. Allmendinger, M. Kaiser, M.A. Itoe, G. Blunden, M.M. Mota, D. Tasdemir. 2013. "Assessment of dual life stage antiplasmodial activity of British seaweeds." *Mar. Drugs* 11: 4019–34.

Stout, E., J. Prudhomme, K. Le Roch, C. Fairchild, S. Franzblau, W. Aalbersberg, M. Hay, J. Kubanek. 2010. "Unusual antimalarial meroditerpenes from tropical red macroalgae." *Bioorganic & Medicinal Chemistry Letters* 20 (19): 5662–65. https://dx.doi.org/10.1016/j.bmcl.2010.08.031.

Sugawara, T., K. Matsubara, R. Akagi, M. Mori, T. Hirata. 2006. "Antiangiogenic activity of brown algae fucoxanthin and its deacetylated product, fucoxanthinol." *J. Agric. Food Chem.* 54 (26): 9805–10. http://doi.org/10.1021/jf062204q.

Sugiura, Y., K. Matsuda, Y. Yamada, M. Nishikawa, K. Shioya, H. Katsuzaki, K. Imai, H. Amano. 2006. "Isolation of a new anti-allergic phlorotannin, phlorofucofuroeckol-B, from an edible brown alga, *Eisenia arborea*." *Biosci. Biotechnol. Biochem.* 70, 2807–11.

Sugiura, Y., Y. Takeuchi, M. Kakinuma, H. Amano. 2006. "Inhibitory effects of seaweeds on histamine release from rat basophile leukemia cells (RBL-2H3)." *Fish. Sci.* 72: 1286–91.

Syad, A., K. Devi. 2015. "Assessment of anti-amyloidogenic activity of marine red alga *G. acerosa* against Alzheimer's beta-amyloid peptide 25-35." *Neurol. Res.* 37: 14–22. http://dx.doi.org/10.1179/1743132814Y.0000000422.

Taskin, E., Z. Caki, M. Ozturk. 2010. "Assessment of in vitro antitumoral and antimicrobial activities of marine algae harvested from the eastern Mediterranean Sea." *Afr. J. Biotechnol.* 9, 4272–77.

Teas, J., 1983. "The dietary intake of *Laminaria*, a brown seaweed, and breast cancer prevention." *Nutr. Cancer.* 4 (3): 217–22.

Teas, J., M. Harbison, R. Gelman. 1984. "Dietary seaweed (*Laminaria*) and mammary carcinogenesis in rats." *Cancer Res.*, 44 (7): 2758–61.

Teas, J., M. Baldeón, D. Chiriboga, J. Davis, A. Sarriés, L. Braverman. 2009. "Could dietary seaweed reverse the metabolic syndrome?" *Asia Pac. J. Clin. Nutr.* 18: 145–54.

Teas, J., T. Hurley, J. Hebert, A. Franke, D. Sepkovic, M. Kurzer. 2009. "Dietary seaweed modifies estrogen and phytoestrogen metabolism in healthy postmenopausal women." *J. Nutr.* 139: 939–44. https://doi.org/10.3945/jn.108.100834.

Teas, J., S. Vena, D. Cone, M. Irhimeh. 2013. "The consumption of seaweed as a protective factor in the etiology of breast cancer: Proof of principle." *J. Appl. Phycol.* 25: 771–79. http://dx.doi.org/10.1007/s10811-012-9931-0.

Topcu, G., Z. Aydoğmuş, S. Imre, A. Gören, J. Pezzuto, J. Clement, D. Kingston. 2003. "Brominated sesquiterpenes from the red alga *Laurencia obtusa*." *J. Nat. Prod.* 66: 1505–8.

Trayhurn, P., I. Wood. 2005. "Signalling role of adipose tissue: adipokines and inflammation in obesity." *Biochem. Soc. Trans.* 33: 1078–81.

Turner, N., J. Thomas, B. Carlson, R. Ogilvie. 1983. "Ethnobotany of the Nitinaht Indians of Vancouver Island." *Occasional Papers of the British Columbia Provincial Museum* 24: 45.

United Nations International Children's Emergency Fund (UNICEF). 2004. "Iodine deficiency still leaves millions of children at risk of mental retardation." *UNICEF*. August 27, 2004.

van Ginneken, V., J. Helsper, W. de Visser, H. van Keulen, W. Brandenburg. 2011. "Polyunsaturated fatty acids in various macroalgal species from north Atlantic and tropical seas." *Lipids Health Dis.* 10: 104. http://dx.doi.org/10.1186/1476-511X-10-104.

Vo, T.-S., D.-H. Ngo, S.-K. Kim. 2012. "Marine algae as a potential pharmaceutical source for anti-allergic therapeutics." *Process Biochem.* 47: 386–94.

Vonthron-Sénécheau, C., M. Kaiser, I. Devambez, A. Vastel, I. Mussio, A.-M. Rusig. 2011. "Antiprotozoal activities of organic extracts from French marine seaweeds." *Mar. Drugs* 9: 922–33.

Wang, J., S. Chen, S. Xu, X. Yu, D. Ma, X. Hu, X. Cao. 2012. "*In vivo* induction of apoptosis by fucoxanthin, a marine carotenoid, associated with down-regulating STAT3/EGFR signaling in sarcoma 180 (S180) xenografts-bearing mice." *Mar. Drugs* 10: 2055–68. http://dx.doi.org/10.3390/md10092055.

Wang, J., W. Jin, W. Zhang, Y. Hou, H. Zhang, Q. Zhang. 2013. "Hypoglycemic property of acidic polysaccharide extracted from *Saccharina japonica* and its potential mechanism." *Carbohydr. Polym.* 95: 143–47. https://doi.org/10.1016/j.carbpol.2013.02.076.

Wang, X., L. Pei, H. Liu, K. Qv, W. Xian, J. Liu, G. Zhang. 2016. "Fucoidan attenuates atherosclerosis in LDLR-/- mice through inhibition of inflammation and oxidative stress." *Int. J. Clin. Exp. Pathol.* 9: 6896–904.

Wang, Y., T. Alexander, T. McAllister. 2009. "*In vitro* effects of phlorotannins from *Ascophyllum nodosum* (brown seaweed) on rumen bacterial populations and fermentation." *J. Sci. Food Agric.* 89 (13): 2252–60.

Weisburger, J., B. Reddy, P. Hill, L. Cohen, E. Wynder, N. Spingarn. 1980. "Nutrition and cancer—on the mechanisms bearing on causes of cancer of the colon, breast, prostate, and stomach." *Bull. N. Y. Acad. Med.* 56: 673–96.

Witvrouw, M., E. De Clercq. 1997. "Sulfated polysaccharides extracted from sea algae as potential antiviral drugs." *Gen. Pharmacol.* 29: 497–511.

Woo, M.-N., S.-M. Jeon, H.-J. Kim, M.-K. Lee, S.-K. Shin, Y.C. Shin, Y.-B. Park, M.-S. Choi. 2010. "Fucoxanthin supplementation improves plasma and hepatic lipid metabolism and blood glucose in high-fat fed C57BL/6N mice." *Chem. Biol. Interact.* 186: 316–22. http://dx.doi.org/10.1016/j.cbi.2010.05.006.

World Health Organization. n.d. "Cardiovascular Diseases." World Health Organization (website). Accessed March 22, 2020. https://www.who.int/health-topics/cardiovascular-diseases/#tab=tab_1.

Wright, A.D., G.M. König, C.K. Angerhofer, P. Greenidge, A. Linden, R. Desqueyroux-Faúndez. 1996. "Antimalarial activity: The search for marine-derived natural products with selective antimalarial activity." *J. Nat. Prod.* 59: 710–16.

Wynder, E., T. Hirayama. 1977. "Comparative epidemiology of cancers of the United States and Japan." *Prev. Med.* 6 (4): 567–94. http://doi.org/10.1016/0091-7435(77)90042-1.

Yamaguchi, M., S. Hachiya, S. Hiratuka, T. Suzuki. 2001. "Effect of marine algae extract on bone calcification in the femoral-metaphyseal tissues of rats: Anabolic effect of *Sargassum horneri*." *J. Health Sci.* 47: 533–38.

Yamamoto, I., T. Nagumo, K. Yagi, H. Tominaga, M. Aoki. 1974. "Antitumor effect of seaweeds, I. Antitumor effect of extracts from *Sargassum* and *Laminaria*." *Japanese J. Exp. Med.* 44: 543–46.

Yamamoto, I., H. Maruyama, M. Takahashi, K. Komiyama. 1986. "The effects of dietary or intraperitoneally injected seaweed preparations on the growth of sarcoma-180 cells subcutaneously implanted into mice." *Cancer Lett.* 30: 125–31.

Yamamoto, K., C. Ishikawa, H. Katano, T. Yasumoto, N. Mori. 2011. "Fucoxanthin and its deacetylated product, fucoxanthinol, induce apoptosis of primary effusion lymphomas." *Cancer Lett.* 300: 225–34. https://doi.org/10.1016/j.canlet.2010.10.016.

Yamasaki-Miyamoto, Y., M. Yamasaki, H. Tachibana, K. Yamada. 2009. "Fucoidan induces apoptosis through activation of caspase-8 on human breast cancer MCF-7 cells." *J. Agric. Food Chem.* 57: 8677–82. http://dx.doi.org/10.1021/jf9010406.

Yamori, Y., A. Miura, K. Taira. 2001. "Implications from and for food cultures for cardiovascular diseases: Japanese food, particularly Okinawan diets." *Asia Pac. J. Clin. Nutr.* 10: 144–45. http://dx.doi.org/10.1111/j.1440-6047.2001.00227.x.

Yan, X.-J., X.-C. Li, X. Fan, C.-X. Zhou. 1997. "Studies on extraction procedure and antioxidative activity of phlorotannins from *Sargassum kjellmanianum*." *Chinese Journal of Oceanology and Limnology* 15: 42–45.

Yan, X., Y. Chuda, M. Suzuki, T. Nagata. 1999. "Fucoxanthin as the major antioxidant in *Hijikia fusiformis*, a common edible seaweed." *Biosci. Biotechnol. Biochem.* 63: 605–7. http://dx.doi.org/10.1271/bbb.63.605.

Yang, E.-J., J.-Y. Moon, M.-J. Kim, D. Kim, W.J. Lee, H.H. Lee, C.-G. Hyun. 2010. "Anti-inflammatory effect of *Petalonia binghamiae* in LPS-induced macrophages is mediated by suppression of iNOS and COX-2." *International Journal of Agriculture and Biology* 12: 754–58.

Yang, J.W., S.Y. Yoon, S.J. Oh, S.K. Kim, K.W. Kang. 2006. "Bifunctional effects of fucoidan on the expression of inducible nitric oxide synthase." *Biochemical and Biophysical Research Communications* 346: 345–50.

Ye, J., Y. Li, K. Teruya, Y. Katakura, A. Ichikawa, H. Eto, M. Hosoi, M. Hosoi, S. Nishimoto, S. Shirahata. 2005. "Enzyme-digested fucoidan extracts derived from seaweed *Mozuku* of *Cladosiphon novae-caledoniae kylin* inhibit invasion and angiogenesis of tumor cells." *Cytotechnology* 47: 117–26. http://dx.doi.org/10.1007/s10616-005-3761-8.

Yokota, T., K. Nomura, M. Nagashima, N. Kamimura. 2016. "Fucoidan alleviates high-fat diet-induced dyslipidemia and atherosclerosis in ApoE^{shl} mice deficient in apolipoprotein E expression." *J. Nutr. Biochem.* 32: 46–54. http://dx.doi.org/10.1016/j.jnutbio.2016.01.011.

Yoon, N., S.-H. Lee, Y. Li, S.-K. Kim. 2009. "Phlorotannins from *Ishige okamurae* and their acetyl- and butyrylcholinesterase inhibitory effects." *J. Funct. Foods* 1 (4): 331–335. http://doi.org/10.1016/j.jff.2009.07.002.

Yoon, W.-J., Y.M. Ham, K.-N. Kim, S.-Y. Park, N.H. Lee, C.-G. Hyun, W.J. Lee. 2009. "Anti-inflammatory activity of brown alga *Dictyota dichotoma* in murine macrophage RAW 264.7 cells." *Journal of Medicinal Plants Research* 3: 1–8.

Yoshida, T., A. Hirano, H. Wada, K. Takahashi, M. Hattori. 2004. "Alginic acid oligosaccharide suppresses Th2 development and IgE production by inducing IL-12 production." *Int. Arch. Allergy Immunol.* 133: 239–47.

Yu, R.-X., X.-M. Hu, S.-Q. Xu, Z.-J. Jiang, W. Yang. 2011. "Effects of fucoxanthin on proliferation and apoptosis in human gastric adenocarcinoma MGC-803 cells via JAK/STAT signal pathway." *Eur. J. Pharmacol.* 657: 10–19. http://dx.doi.org/10.1016/j.ejphar.2010.12.006.

Yuan, Y.V., M. Carrington, N. Walsh. 2005. "Extracts from dulse (*Palmaria palmata*) are effective antioxidants and inhibitors of cell proliferation in vitro." *Food Chem. Toxicol.* 43 (7): 1073–81.

Yuan, Y.V., N.A. Walsh. 2006. "Antioxidant and antiproliferative activities of extracts from a variety of edible seaweeds." *Food Chem. Toxicol. Int. J. Publ. Br. Ind. Biol. Res. Assoc.* 44: 1144–50. http://dx.doi.org/10.1016/j.fct.2006.02.002.

Zaneveld, J.S. 1959. "The utilization of marine algae in tropical south and east Asia." *Econ. Bot.* 13: 89–131.

Zhang, C., Y. Li, Z.-J. Qian, S.-H. Lee, Y.-X. Li, S.-K. Kim. 2011. "Dieckol from *Ecklonia cava* regulates invasion of human fibrosarcoma cells and modulates MMP-2 and MMP-9 expression via NF-κB pathway." *Evid. Based Complement. Alternat. Med.* 2011. https://doi.org/10.1155/2011/140462.

Zhang, Z., P. Zhang, M. Hamada, S. Takahashi, G. Xing, J. Liu, N. Sugiura. 2008. "Potential chemoprevention effect of dietary fucoxanthin on urinary bladder cancer EJ-1 cell line." *Oncol. Rep.* 20: 1099–103.

Zhang, Z., Q. Zhang, J. Wang, H. Song, H. Zhang, X. Niu. 2010. "Chemical modification and influence of function groups on the in vitro-antioxidant activities of porphyran from *Porphyra haitanensis*." *Carbohydr. Polym.* 79: 290–95. http://dx.doi.org/10.1016/j.carbpol.2009.08.004.

Zhang, Z., K. Teruya, H. Eto, S. Shirahata. 2013. "Induction of apoptosis by low-molecular-weight fucoidan through calcium- and caspase-dependent mitochondrial pathways in MDA-MB-231 breast cancer cells." *Biosci. Biotechnol. Biochem.* 77: 235–42. http://dx.doi.org/10.1271/bbb.120631.

Zou, Y., Z.-J. Qian, Y. Li, M.-M. Kim, S.-H. Lee, S.-K. Kim. 2008. "Antioxidant effects of phlorotannins isolated from *Ishige okamurae* in free radical mediated oxidative systems." *Journal of Agricultural and Food Chemistry* 56: 7001–9.

Algae: A Global Perspective

Anusonadisai, Nattasuda. 2015. "Slavery in shrimp industry known to Thailand government." *CBC News*. December 21, 2015. https://www.cbc.ca/news/business/slavery-shrimp-thailand-1.3374433.

Baker, R. 2020. "B.C. First Nations Leaders Want Immediate End to Open-Net Salmon Farms." *National Observer*. June 30, 2020. https://www.nationalobserver.com/2020/06/30/news/bc-first-nations-leaders-want-immediate-end-open-net-salmon-farms.

Consumer Reports. 2015. "How safe is your shrimp? *Consumer Reports*' guide to choosing the healthiest, tastiest, and most responsibly sourced shrimp." April 24, 2015. https://www.consumerreports.org/cro/magazine/2015/06/shrimp-safety/index.htm.

Davies, R., S. Cripps, A. Nickson, G. Porter. 2009. "Defining and estimating global marine fisheries bycatch." *Marine Policy* 33 (4): 661–72. https://doi.org/10.1016/j.marpol.2009.01.003.

Department of Fisheries and Oceans Canada. n.d. "Farmed Salmon." Department of Fisheries and Oceans Canada (website). Last modified March 15, 2017. https://www.dfo-mpo.gc.ca/aquaculture/sector-secteur/species-especes/salmon-saumon-eng.htm.

Easton, M., D. Luszniak, E. Von der Geest. 2002. "Preliminary examination of contaminant loadings in farmed salmon, wild salmon and commercial salmon feed." *Chemosphere* 46: 1053–74. https://doi.org/10.1016/S0045-6535(01)00136-9.

Environmental Justice Foundation. 2003. *Smash & Grab: Conflict, Corruption and Human Rights Abuses in the Shrimp Farming Industry*. https://ejfoundation.org/resources/downloads/smash_and_grab.pdf.

Findlay, Andrew. 2018. "A Deep-Dive into the controversy of Salmon Farms in British Columbia." *Douglas*. April 21, 2018. https://douglasmagazine.com/controversy-salmon-fish-farms-british-columbia/.

Food and Agriculture Organization of the United Nations. n.d. "Farmed shrimp output increased by about 6 percent in 2016." *GLOBEFISH - Information and Analysis on World Fish Trade*. Accessed May 30, 2020. http://www.fao.org/in-action/globefish/market-reports/resource-detail/en/c/1136583/.

Gerber, P., H. Mooney, J. Dijkman, S. Tarawali, C. de Haan. 2010. *Livestock in a Changing Landscape*. Vol. 2, *Experiences and Regional Perspectives*. Island Press.

Govender, R., K. Hayne, S.D. Fuller, S. Wallace. 2016. *Taking Stock: Sustainable Seaweed in Canadian Markets*. SeaChoice. http://www.seachoice.org/wp-content/uploads/2016/06/Seachoice-Taking-Stock-Report-June-7.pdf.

Guy, Allison. 2017. "5 Facts That Will Make You Think Twice About Eating Imported, Farm-Raised Shrimp." *Oceana*. February 14, 2017. https://oceana.org/blog/5-facts-will-make-you-think-twice-about-eating-imported-farm-raised-shrimp.

Hawken, Paul. 2017. *Drawdown: The Most Comprehensive Plan Ever Proposed to Reverse Global Warming*. Penguin Random House.

Henry, B., K. Laitala, I. Klepp. 2019. "Microfibres from apparel and home textiles: Prospects for including microplastics in environmental sustainability assessment." *Sci. Total Environ.* 652: 483–94. https://doi.org/10.1016/j.scitotenv.2018.10.166.

Hites, R., J. Foran, D. Carpenter, M. Hamilton, B. Knuth, S. Schwager. 2004. "Global assessment of organic contaminants in farmed salmon." *Science* 303: 226–29. http://doi.org/10.1126/science.1091447.

Kinley, R., R. de Nys, M. Vucko, L. Machado, N. Tomkins. 2016. "The red macroalgae *Asparagopsis taxiformis* is a potent natural antimethanogenic that reduces methane production during *in vitro* fermentation with rumen fluid." *Animal Prod. Sci.* 56: 282–89. http://dx.doi.org/10.1071/AN15576.

Kolanjinathan, K., P. Ganesh, M. Govindarajan. 2009. "Antibacterial activity of ethanol extracts of seaweeds against fish bacterial pathogens." *European Review for Medical and Pharmacological Sciences* 13: 173–77.

Kole, P., A. Löhr, F. Van Belleghem, A.M.J. Ragas. 2017. "Wear and tear of tyres: A stealthy source of microplastics in the environment." *Int. J. Environ. Res. Public Health* 14: 1265.

Kosuth, M., S. Mason, E. Wattenberg. 2018. "Anthropogenic contamination of tap water, beer, and sea salt." *PLOS One* 13 (4): e0194970. https://doi.org/10.1371/journal.pone.0194970.

Laville, Sandra, Matthew Taylor. 2017. "A million bottles a minute: World's plastic binge 'as dangerous as climate change.'" *The Guardian*. June 28, 2017. https://www.theguardian.com/environment/2017/jun/28/a-million-a-minute-worlds-plastic-bottle-binge-as-dangerous-as-climate-change.

Lebreton, L., B. Slat, F. Ferrari, B. Sainte-Rose, J. Aitken, R. Marthouse, S. Hajbane et al. 2018. "Evidence that the Great Pacific Garbage Patch is rapidly accumulating plastic." *Sci. Reports* 8: 4666 (2018). http://dx.doi.org/10.1038/s41598-018-22939-w.

Li, X., H. Norman, R. Kinley, M. Laurence, M. Wilmot, H. Bender, R. de Nys, N. Tomkins. 2016. "*Asparagopsis taxiformis* decreases enteric methane production from sheep." *Animal Prod. Sci.* 58: 681–88. http://dx.doi.org/10.1071/AN15883.

Mason, S., V. Welch, J. Neratko. 2018. "Synthetic polymer contamination in bottled water." *Front Chem.* 6: 407. https://doi.org/10.3389/fchem.2018.00407.

Mato, Y., T. Isobe, H. Takada, H. Kanehiro, C. Ohtake, T. Kaminuma. 2001. "Plastic resin pellets as a transport medium for toxic chemicals in the marine environment." *Environ. Sci. Technol.* 35: 318–24. https://doi.org/10.1021/es0010489.

JESSE SWINIMER PHOTO

Morton, A. 2020. *Sea Lice Survey Four Regions of BC Coast 2020: Preliminary Report*. June 16, 2020. https://d3n8a8pro7vhmx.cloudfront.net/safesalmon/pages/69/attachments/original/1597275993/Sea_Lice_Survey_Four_Regions_of_BC_Coast_2020sm_%281%29.pdf?1597275993.

Morton, A., R. Routledge, S. Hrushowy, M. Kibenge, F. Kibenge. 2017. "The effect of exposure to farmed salmon on piscine orthoreovirus infection and fitness in wild Pacific salmon in British Columbia, Canada." *PLOS One* 12 (12): e0188793. http://dx.doi.org/10.1371/journal.pone.0188793.

Napper, I., R. Thompson. 2016. "Release of synthetic microplastic plastic fibres from domestic washing machines: effects of fabric type and washing conditions." *Marine Pollution Bulletin* 112 (1): 39–45. https://doi.org/10.1016/j.marpolbul.2016.09.025.

Office for Coastal Management. n.d. "Fast Facts: Coral Reefs." National Oceanic and Atmospheric Administration (website). Accessed March 15, 2019. https://coast.noaa.gov/states/fast-facts/coral-reefs.html.

Pacific Salmon Foundation. n.d. "Salmon Facts." Pacific Salmon Foundation (website). Accessed June 3, 2020. https://www.psf.ca/learn/salmon-facts.

Research and Markets. 2018. "Global Bottled Water Market 2018–2023: A $334 Billion Opportunity, Driven by Health & Wellness Concerns." Business Wire (website). August 13, 2018. https://www.businesswire.com/news/home/20180813005378/en/Global-Bottled-Water-Market-2018-2023-334-Billion.

Roque, B., C. Brooke, J. Ladau, T. Polley, L. Marsh, N. Najafi, P. Pandey et al. 2019. "Effect of the macroalgae *Asparagopsis taxiformis* on methane production and rumen microbiome assemblage." *Animal Microbiome* 1: 3 (2019). https://doi.org/10.1186/s42523-019-0004-4.

Sandvik, Kjersti, Aud Slettehaug, Liudmilla Ilina. 2018. "Salmon treated like chicken." Forum, Terra Madre Salone del Gusto. September 20, 2018. Turin, Italy.

SeaChoice. 2018. "Canadian Imports/Exports." SeaChoice (website). February 13, 2018. https://www.seachoice.org/info-centre/markets/canadian-imports-exports/.

Slow Fish. n.d. "Overfishing." Slow Food (website). Accessed June 18, 2020. https://www.slowfood.com/what-we-do/themes/slow-fish/threats-to-the-ocean/overfishing/.

Smart, Amy. 2017. "B.C. fish farms: A tangled net." *Times Colonist*. December 3, 2017. https://www.timescolonist.com/islander/b-c-fish-farms-a-tangled-net-1.23111384.

Sprague, M., J. Dick, D. Tocher. 2016. "Impact of sustainable feeds on omega-3 long-chain fatty acid levels in farmed Atlantic salmon, 2006–2015." *Scientific Reports* 6: 21892. https://doi.org/10.1038/srep21892.

Stokstad, Erik. 2012. "The carbon footprint of a shrimp cocktail." *Science*. February 17, 2012. https://www.sciencemag.org/news/2012/02/carbon-footprint-shrimp-cocktail.

The Ocean Cleanup. Accessed April 1, 2019. The Ocean Cleanup (website). https://theoceancleanup.com.

Thomson, Janet, Manmeet Ahluwalia. 2012. "Remembering the mighty cod fishery 20 years after moratorium." *CBC*. June 29, 2012. https://www.cbc.ca/news/canada/remembering-the-mighty-cod-fishery-20-years-after-moratorium-1.1214172.

United Nations Educational, Scientific and Cultural Organization (UNESCO). n.d. "World Oceans Day." UNESCO (website). Accessed March 24, 2019. http://www.unesco.org/new/en/unesco/events/prizes-and-celebrations/celebrations/international-days/world-oceans-day-2014/.

World Counts. n.d. "Tons of meat eaten." Accessed May 19, 2021. https://www.theworldcounts.com/challenges/consumption/foods-and-beverages/world-consumption-of-meat/story.

World Health Organization. 2019. *Microplastics in Drinking-Water*. https://apps.who.int/iris/bitstream/handle/10665/326499/9789241516198-eng.pdf?ua=1.

Zhang, L., I. Rana, R. Shaffer, E. Taioli, L. Sheppard. 2019. "Exposure to glyphosate-based herbicides and risk for non-Hodgkin lymphoma: A meta-analysis and supporting evidence." *Mutation Research/Reviews in Mutation Research* 781: 186–206. https://doi.org/10.1016/j.mrrev.2019.02.001.

INDEX

Page numbers in **bold** include photos or illustrations

My good friend and artist Heather Good created this piece for my birthday present one year. HEATHER GOOD PAINTING

ABOUT THE AUTHOR

Amanda Swinimer (BSc) lives on the west coast of Vancouver Island, BC, with her two daughters. She has her own business, Dakini Tidal Wilds, hand harvesting wild seaweed from the beaches and kelp forests surrounding her home. Swinimer's unique expertise makes her a sought-after speaker at international conferences about the ecological and environmental importance of seaweeds. She has been conducting seaweed tours as well as teaching young people in the British Columbia school system for years, passing on knowledge about the health benefits of seaweeds and sharing her intensely joyful connection to the ocean with diverse audiences. Her first love is her responsibly harvested seaweed, which is sold to some of British Columbia's finest restaurants and available through her website.

Amanda and her daughters, Mahina and Nesika, on nori-covered rocks at the beach in front of their home.